建筑立场系列丛书 No.68

休养和度假住宅

Retreats and Escapes

OFIS arhitekti建筑师事务所等 | 编

倪琪 栾一斐 | 译

大连理工大学出版社

C3 建筑立场系列丛书 No. 68

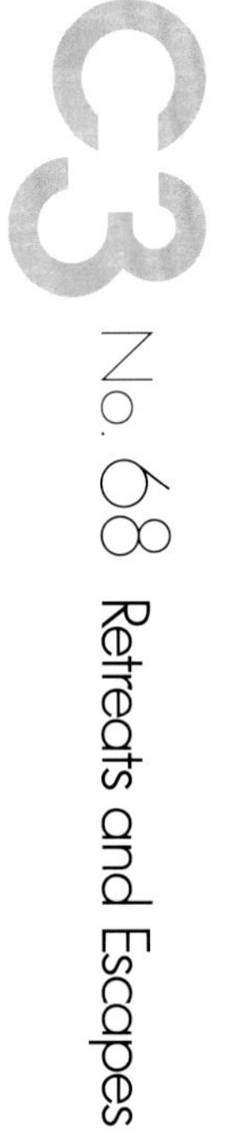

Retreats and Escapes

Meditation

Mille Arbres——一个新型多用途项目 _Sou Fujimoto Architects + Manal Rachdi OXO Architects

Mille Arbres，或称"千树"，是一个新型的多用途项目，坐落于马约门附近的潘兴广场，地处连接着巴黎市中心与拉德芳斯的历史中轴线之上。

受奥斯曼建筑风格的影响，Mille Arbres成了巴黎的新生活方式。

这个项目选择把住宅放在了建筑的顶部，将地面区域留出来规划为城市公园。建筑的倒金字塔形式由于尽量减少了占地面积，所以提供了最大化的公共区域和底层绿化区域。公园的设计将会是保证整个地区健康生活的重要元素，也将成为连接巴黎与市郊的纽带。

项目包含办公区域、酒店、美食街、多种多样的儿童设施以及公交总站。

本建筑采用渗透式设计，多条道路贯穿其中。为了不影响下部设施的功能，这个城市公园项目在一些位置上被升高以保持连贯性。底层的商业内街和美食广场成了此项目的核心元素。

顺着升高的地势往上走，可以看到"生物多样性"走廊，在这里可以体验到如真正森林一样的生态系统。大型圆形露天剧场连接着走廊与道路，这里可以举行会议和文化活动。

倒金字塔中包含了酒店和办公室。它的独特体量与场地前方的议事厅的规模遥相呼应。大型露台为建筑中央带来了自然光，为整个建筑内部提供了多元化的风景，让住户可以看到街面和天空的景色。楼顶两层共有127个居住单元。它们有着相似的建筑风格，成为世界上首个由生物质材料建造的屋顶村落。这些住房的周围设置有人行小路，与大自然和谐相融，在这里人们随时能听见风吹树叶的沙沙声，阳光透过树叶的缝隙洒在房间里，为住户揭开了地平线、巴黎屋顶和埃菲尔铁塔的神秘面纱。在这个环形公路的上空，人们犹如置身于茂密的森林一般。

"千树"开发了一种策略，它整合了这个场地的矛盾条件，与之发生相互作用，并做出了优化处理，使混乱性与复杂性的特点相结合。这个策略为项目提供了可逆性和灵活性，也是当今社会所面临的主要挑战。在该项目中，每个房间都可以被用作酒店、办公室或者住宅。随着时间的推移，这个项目将会适应城市的各种改造。

Mille Arbres, a New Mixed-use Project

Mille Arbres, or a Thousand Trees, is a new mixed-use project located on the Pershing site near Porte Maillot, on the historical axis linking the heart of Paris to La Defense.
Inspired by the Hausmannian block building typology, Mille Arbres is a new way of living in Paris.
The project offers to raise the residential program to the top of the building, leaving the ground free to serve as an urban park. The reversed pyramid shape of the structure maximizes the public space and the planted areas on the ground floor by having a minimal footprint. The park created will become an important element for the well-being of the district as well as a soft connection between Paris and its outskirts.
The program includes offices spaces, a hotel, an inner street with a food court, various programs for children and a bus terminal.
The building is designed to be very permeable, with multiple pathways through it. The topography of the urban park is lifted in places to facilitate the connections while allowing for programs underneath. The inner street on the ground level (La rue Gourmande), together with its food court become core elements of the project.
Following the lifted topography, there is the "bio-diversity" path, where you can experience the real ecosystem of a forest. The paths connect back to the street level through large amphitheaters for meetings and cultural events.
The reversed pyramid houses the hotel and offices. Its unique volume responds to the scale of the congress hall in front of the site. Large patios bring natural light into the heart of the building, offering a multiplicity of views within the entire project and connecting the occupants to the street and the sky.
127 residential units exist on the top two floors. They are designed with similar architectural qualities to create on the top floor the first village built out of bio-based materials in the world. Organized around pedestrian paths, the houses merge with nature, where the wind blows through the trees and the sunlight permeating through the leaves and unveiling a surreptitious view to the horizon; The roofs of Paris and the Eiffel Tower. On the ring-road, we live in a forest.
Mille Arbres develops a strategy that integrates, reacts and optimizes the contradictive conditions of the site, combining promiscuity and complexity. It offers reversibility and flexibility for the programs which are the major challenges of our generation. We designed a project that allows the offices to become housing hotel, office or residential buildings. The project will be able to adapt to the city's transformations over time.

café restaurant

In Vivo——真正的实验大楼 _ BPD Marignan + XTU Architects + SNI Group + MU Architecture

该项目是由BPD Marignan与XTU建筑师事务所、SNI设计集团以及MU建筑设计公司共同完成的，获得了以巴黎塞纳河左岸的景观改造为目标的设计竞赛的第一名。

该项目将建造多栋旗舰大楼，为城市物质生活与自然环境的整合提供参考依据。这座真正的实验大楼坐落于巴黎13街区的元帅大道上，以其独特的活性生物立面设计为特色，在立面上进行微藻培养以供医学研究。

鉴于不断严重的过度城市化、化石燃料的枯竭以及日益明显的不平等状态对环境和健康的影响，许多城市都面临着一项21世纪的主要问题：促进社会融合，也推动城市居民之间的开放性，让大自然融入城市景观，从而实现一个更加公平、更加有弹性的可持续发展的城市。

第一步，In Vivo项目通过创建一个经过建筑规划的环境来回应了这个问题，它促进了社会性和功能性的融合，使聚会与分享活动发生在楼内居民、使用者、邻里之间及至整个巴黎市的市民之间。

该项目包括三栋大楼，它们都注重以下方面的设计：自然风光、曝光度、照明、通风以及与城市的最佳结合方式。

In Vivo项目立志成为自然与社会相结合的创新典范，除了三栋供人类居住的大楼之外，还有一栋特殊建筑中装有蚯蚓。

——绿树住宅：各个立面都安装有大型阳台花箱，种有茂盛的绿树和灌木丛，保证城市的生物多样性

——植被住宅：在温室下面或是室外房顶上有各种各样的菜园和小型都市农业凉亭

——藻类住宅：整合安装了培养微藻类的立面用于医学研究

——蚯蚓大楼：饲养蚯蚓，用于对住户的生物垃圾进行蚯蚓堆肥和条件培养

In Vivo, Genuine Laboratory Buildings

BPD Marignan and XTU Architects, associated with SNI Group and MU Architecture have won the Réinventer Paris competition for Paris Rive Gauche site M5A2.

They will put up flagship buildings, clear proof of the integration of living matter and nature in the city. This genuine laboratory-building located on Boulevard des Maréchaux in the 13th district of Paris, will feature an active biofacade housing microalgae culture for medical research.

In view of the environmental and health impacts due to the soaring hyper-urbanization, the ending of fossil resources and rising inequalities, cities are facing a major issue of the 21st

century: promoting social mix and openness between citizens and integrating nature into cities, to achieve a fairer, more sustainable and resilient society.

As a first step, the In Vivo project addresses this issue by creating an architectural and programming environment to foster porosity, social and functional mix, meetings and sharing among inhabitants, users, neighbours and Parisians of the Grand Paris.

It has created three different buildings focusing on natural views, exposures, lighting and ventilation, as well as an optimal urban integration.

The In Vivo project is meant to be a manifesto for the innovative urban integration of nature and living matter with three buildings for humans and one for earthworms.

- The Tree House: growing trees and bushes suited to host urban biodiversity on large flower box balconies on all its facades
- The Plant House: dedicated to all forms of vegetable gardens and small-scale urban agriculture in loggias, under greenhouses or outdoor on rooftops
- The Algo House: integrating a microalgae-producing biofacade for medical research
- The Lombric House: raising earthworms to allow the vermicomposting of inhabitants' organic waste and culture conditioning

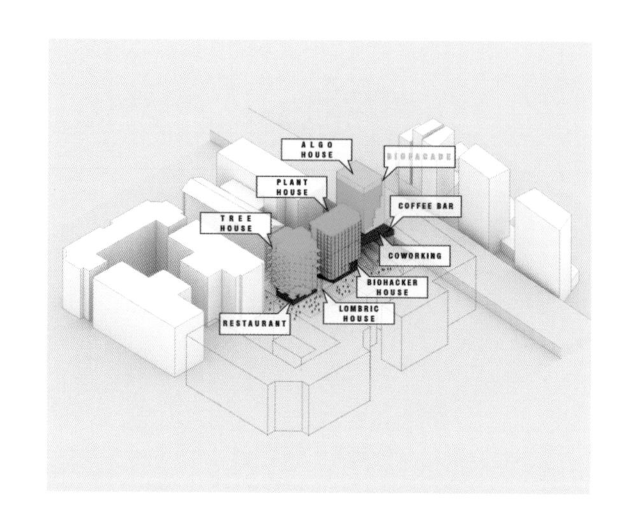

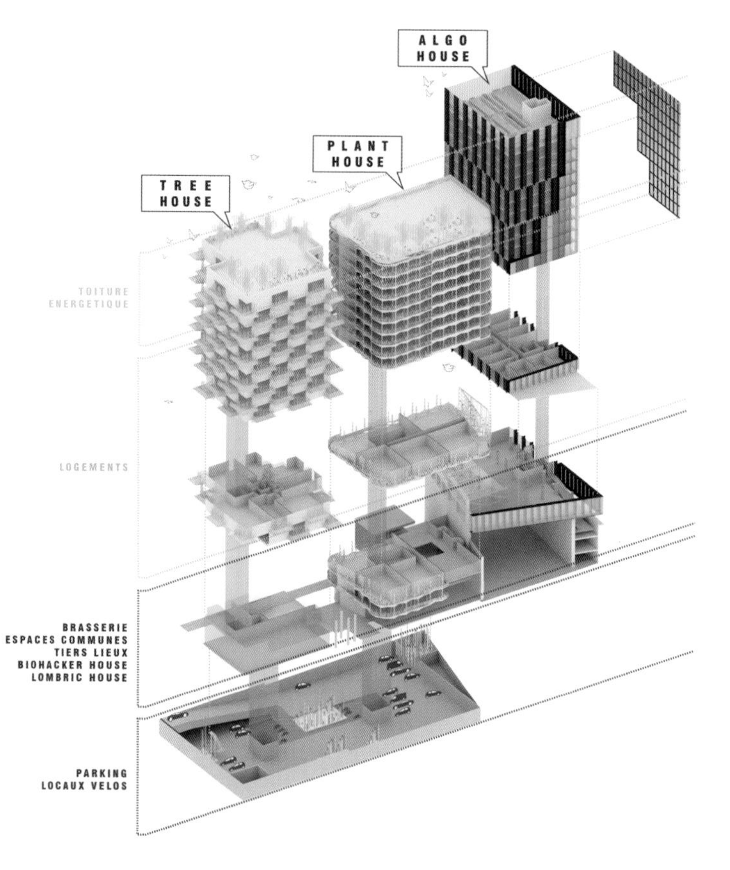

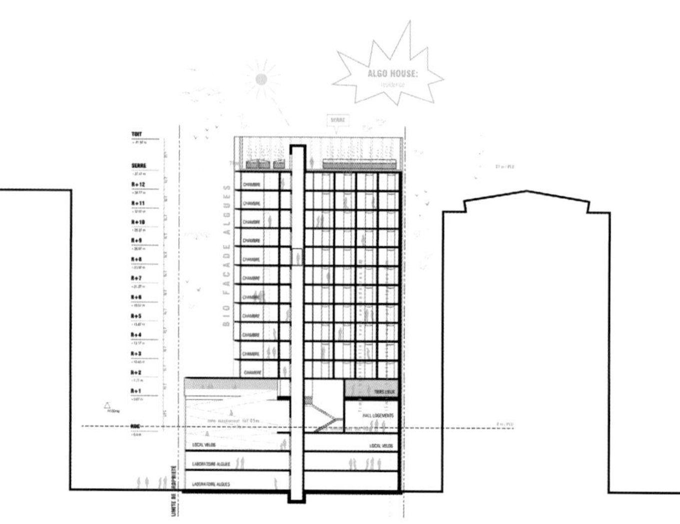

A–A' 剖面图 section A-A'

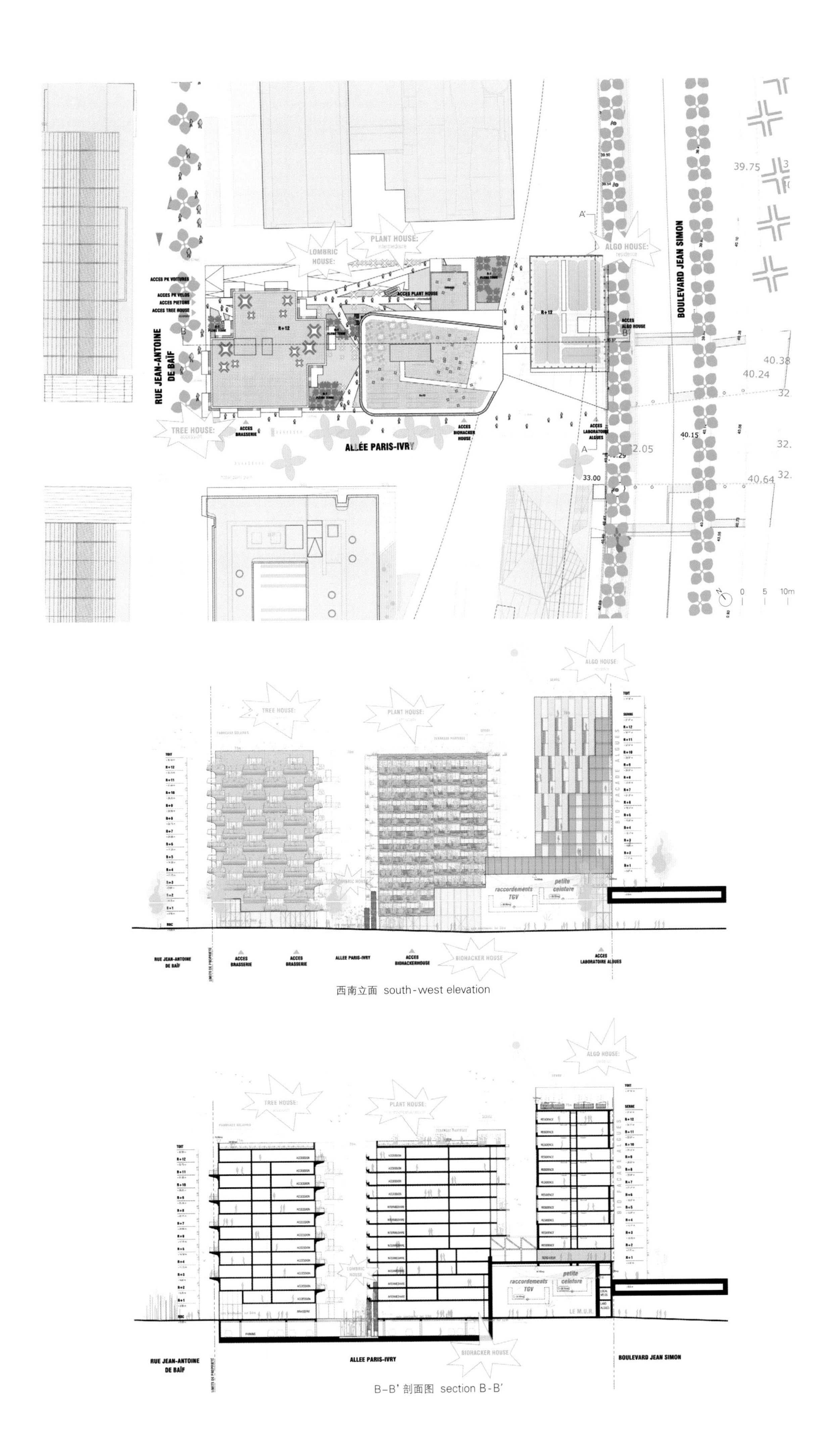

西南立面 south-west elevation

B-B' 剖面图 section B-B'

休养和度假住宅

Retreats and

最近的威尼斯双年展对当代建筑实践中的道德问题展开了讨论。跟以往一样，其中一个最主要的问题就是住房问题。在当前的世界里，难民危机、临建棚营地问题和发达经济社会的住房危机、未来住房问题比比皆是，从政治角度而言，建筑设计在这里的作用至关重要。

在不同的时期，建筑设计都在提出关于居住意义的各种挑战性问题。20世纪住房供应实现了工业化，但是从根本上来说，房屋仍然是一个固定的占有一定空间的物体，其所有的意图和目的都是成为一个永久的人文景观。尽管有人做出了一些尝试来削减它的永久性，但住房仍然是一个非常有持久性的政治课题。

但是，有些建筑却不需要提供永久居住的空间，这里不常住人，如有需要可以拆走在别处重新安装。有一些类似于住宅的空间，像酒店、招待所、小屋和度假屋，它们不断变化的建筑风格以及随着时间变化如何进行改造，都给人留下了深刻的印象。

The recent Venice Biennale attempted to ask ethical questions about contemporary architectural practice, and as ever, one of the main problems is that of housing. In a world of refugee crises, informal settlements and housing crises in the advanced economies, the future of housing, and architecture's role in it, is politically vital.

At different times, architecture has asked challenging questions about what it means to live, or to dwell. The twentieth century industrialised the provision of housing, but fundamentally the house remains a stationary, place-held object, for all intents and purposes an eternal part of the human landscape. Attempts have been made to dislodge this permanence, but the home is a remarkably enduring political object.

But architecture is frequently called upon to create living spaces that are not permanent, that are inhabited infrequently, or that can be moved and relocated if necessary. Spaces at the edge of dwelling, such as hotels, hostels, cabins and retreats, can give an impression of how the architecture of the home is in flux, and how it might change over time.

Escapes

居住形式的多样化

人类大部分是定居生物。在我们悠久的历史上，在各个不同的地区都曾出现过流动居住的民族，他们或是游牧民族或是商旅人士；而在当今社会，也有全球性流动的精英阶层和遍布各地的悲惨难民。但是世界上大多数人还是期望有一个固定的居所，我们称之为“家”。

关于家或是住所的问题，一直是工业世界人类面临的一个重要问题。在第二次世界大战后雄心勃勃的时代，政府修建了成千上万的科技型住宅。世界各地的设计师都设想在不远的将来，家将成为一个短暂的居所，可以随心所欲被移动、更换，甚至加以改造。

哲学家赞赏游牧民的想法，对现代消费社会的人群可以获得自由而感到兴奋。他们认为，自由在生活中各个部分的根本性扩张会打破这个时代的社会和政治局限性。

但另一方面，保守派坚称，人类总会与特定的地点产生关联，有一种纽带将人类和他们的住所联系在一起。其代表是海德格尔关于德国黑森林一间农舍的著名探讨，只有与那个地点有着深刻而亲密的关联，才能拥有更加真实的人生经历。

最近，住房危机和难民危机将替代型住房的设计又提上日程，它是否适合临建棚营地使用？智利建筑师亚力杭德罗·阿拉维纳的“半建成”房屋、公共住宅模板和一系列的试验方法也都投入研究，为市场底层的千禧一代提供帮助。

在最近的威尼斯双年展上，这些问题比较突出，但是与非家用的住宅共同呈现出来，非家用住宅在“分享”和“体验”的新经济形式的影响下同样经历了改变。它们包括酒店、度假屋和其他临时居所。随着时间的迁移，它们的建筑风格的变革紧随住宅风格变革的脚步。

Varieties of Dwelling

Humans are, by and large, sedentary creatures. In our deep history and in various locations we have been nomadic, as hunter-gatherers or trading peoples, while in today's world we have a globe-spanning elite class, and a tragically expanding refugee population. But the vast majority of people in the world expect to have a single location that we can call "home".

The question of home, or of dwelling, is one that has been very important to humans in an industrial world. In the ambitious era after the Second World War, governments built many millions of technological dwellings, and architects all over the world imagined a near future in which the home as an object would be developed into an ephemeral dwelling unit, able to be moved, replaced, or altered with great ease.

Philosophers of the time eulogised the idea of the nomad - excited by the personal freedoms that modern consumer society was making possible, they thought a radical extension of these freedoms to all parts of life would transcend the era's social and political limitations.

But on the other hand, a conservative type of thought asserts that human beings are always going to be connected to specific locations, that there is an existential bond that draws humans to places in which they can dwell. This is epitomised by Heidegger's famous discussion of a peasant's farmhouse in the Black Forest of Germany, that through its deep and intimate connection to its location, allowed for a more authentic kind of human experience.

Recently, housing and refugee crises have brought alternative modes of dwelling back onto the agenda, whether it's interest in informal settlements, Alejandro Aravena's "half-built" houses, co-housing models, or a number of experimental approaches being developed for "millenials" at the bottom of the market.

图片提供：©Anders Hviid

VIPP庇护所，瑞典
The VIPP Shelter, Sweden

"精品酒店"不是一个新概念，而是需要在住房市场着手解决的问题，通过股份制经济和提高服务质量，比如房屋租赁社区Airbnb网站，独特性只会随时间的变化而增值，而建筑设计为这种多样化的感觉做出了主要贡献。

在泰国的大城府，曼谷建筑师事务所Onion设计了一家名为"大城府撒拉酒店"的度假酒店。它非常小，只有26间客房，它的精品品质体现在将餐厅和走廊交互排列。客房本身很宽敞，室内设计拥有spa一样的奢华效果，以白色为主色调，每间客房都有大浴盆。

酒店建筑地处高处，以防洪水灾害，建筑师在许多环境下都采用了薄薄的泥土砖，将房间与泳池的优雅特色一同展现出来；尤其还用在一个起到通道作用的院子里，其平面布局为圆齿形曲线形式，形成了不断变化的光影交替。这既是一种熟悉的奢侈建筑，又是一个以建筑设计为主导营造的环境。

Tavaru餐厅酒吧位于马尔代夫一个小岛上的私人度假中心，由总部位于布拉格的ADR建筑师事务所设计。它体现了"体验文化"在休闲住宿设计中的重要性。

与周围的建筑相比，该建筑明显呈垂直状，由许多小型的圆形楼面板垂直堆放而成。这座建筑有两个功能——一个两层楼高的封闭葡萄酒酒窖，顺着楼梯往上走有一家美食餐馆。整座建筑都被包裹在半透明的织物里，外观看起来与众不同。

这个小项目专注于奢华而完全脱离了现实生活，与家的概念毫无瓜葛，建筑设计成了体验华丽改造的一部分。

几乎没有其他的工业像酿酒业一样兴起了"体验"这一概念。在这个竞争越来越激烈的世界市场中，随着新地区的不断加入，利润空

These issues were prominent at the recent Venice Biennale, but take their place alongside more established forms of non-homes that also are experiencing change due to the new economics of "sharing" and "experience". These include hotels, retreats, and other temporary types of accommodation. Their own architectural evolution points at ways in which housing itself is changing over time.

The "boutique hotel" is not a new concept, but in an accommodation market assailed by the sharing economy and the rise of services such as Airbnb, uniqueness is something that is only going to increase in value over time, and architecture is a major contributor to this sense of variety.

In the Phra Nakhon Si Ayutthaya district of Thailand, Bangkok architects Onion designed a resort hotel called Sala Ayutthaya. With only twenty-six rooms, it is very small, and its boutique character is enhanced by the cross-programming of a restaurant and gallery space into the sequence. The rooms themselves are large, and the interiors have been designed with a spa-like luxury effect - plenty of white, large baths in every room.

Lifting the building up to avoid the risk of floods, the architects have juxtaposed the ethereal qualities of the rooms and the pool with the use of a thin, earthy brick in a number of contexts – in particular a courtyard used for circulation, where scalloped curves in plan create a constantly shifting play of light and shadow. This is both a familiar kind of luxury building, but one in which the architecture plays a leading role in the creation of atmosphere.

The Tavaru Restaurant and Bar is the centerpiece of a private resort on a tiny island in the Maldives, and is the work of Prague-based ADR architects. It is an example of the importance that the culture of "experience" is becoming to leisure and accommodation design.

In contrast to its surroundings, the building is expressly vertical, with very small circular floor plates stacked vertically. The building has two functions - a vintage wine cellar is a closed double-height volume, while stairs lead up to a gastronomic restaurant above. The whole building is wrapped in translucent

“安托万”庇护所，瑞士
Antoine, Switzerland

斯古塔高山庇护所，斯洛文尼亚
Alpine Shelter Skuta, Slovenia

间变得越来越小。近几年，许多葡萄园都引入了顶级建筑师来设计游客中心和酒店，并通过其他一些方式利用他们的品牌开辟新的收入来源。

当人们想象未来住房模式的时候，其中一个最经常提到的主题就是个人庇护所的概念：如果每个人的家都是独立的、量身定制的、与众不同的，城市会变成什么样？大城市中非正式的贫民窟为这种想法提供了样板，但是农村住房、远离尘嚣的单体建筑提供了另一种样板。

丹麦VIPP住房设计公司设计了一座特殊的小屋作为他们产品的样品。它购买方便，可以安装在任意位置，这样看起来，它更像是一个工业产品而非树林里的传统小屋。

它由VIPP的首席设计师Morten Bo Jensen设计，是一个预制的钢质箱子，两边装有玻璃，安装在支柱上以适应地势。这家公司善于制造高品质的钢材产品，黑色的表皮与炭灰色的各种内部材料相得益彰——钢质家具、厨房台面和粗糙的墙面。

不愧是从事工业产品制造的，小屋的上层结构刚好可以容纳一张单人床，位于屋顶天窗下方，形成了一种完全独立的、私人化的、对自然环境的深刻体验。

斯洛文尼亚OFIS建筑师事务所的建筑师与哈佛大学的学生将目光放在极端气候下的庇护所上。这个庇护所是为阿尔卑斯山上的一处露营地设计的，新的庇护所对于山上的人来说生死攸关。

庇护所采用钢结构，木质外壳和内部结构，外部覆盖薄混凝土嵌板。它可被划分为三个不同的单元，每一个单元均为陡坡屋顶，防止冬天积雪。这三个单元大部分都是提前在工厂建好的，然后用直升飞机运送到指定位置。

fabric, giving the building a strikingly alien appearance.
This tiny project is entirely devoted to a very rarified concept of luxury, completely detached from everyday life - far away from home, architecture becomes part of a spectacular transformation of experiences.
Almost no other industry has embraced the “experience” concept more than that of winemaking. In a more and more competitive world market, with new regions entering constantly, profit is squeezed. A great many vineyards have brought in top-class architects over recent years, to create visitors' centers, hotels and other ways to use their brands to open new sources of revenue.

When people imagine the future of housing, one of the themes that frequently comes up is the idea of the personal shelter - what would the city be like if everyone's home was individual, customized, different? The informal slums of the great cities provide one model for this kind of thinking, but the rural shelter, single dwellings set off from the world, provide another example.
Danish design house VIPP, has designed a specialised cabin as a demonstration of their products. It is available to purchase and be installed anywhere, and in this way is more like an industrial product than a traditional hut in the woods.
Designed by Morten Bo Jensen, VIPP's chief designer, the shelter is a prefabricated steel box, glazed on both sides, and installed on stilts to adapt to the terrain. Demonstrating the quality steel work that is the company's specialty, its dark skin is matched on the interior by a variety of materials in the same charcoal grey - steel furniture, kitchen surfaces, and a matte felt for the walls.
Unashamedly industrial, the cabin features an upper level just large enough for a double bed, positioned under a rooflight, an intimate, yet fully separate, branded experience of natural surroundings.
Led by Slovenian architects OFIS with students from Harvard, looks at shelter in an extreme climate. Designed for a site high up in the Slovenian Alps, the new shelter is in a location that

种马住宅，美国华盛顿
Studhorse, Washington, USA

庇护所所处的环境可能景色如画，但是将这个建筑放置在这一位置，使此庇护所彻底融入了游牧民族的庇护所风格——让人想起了用直升机运送巴克敏斯特·富勒的网格状球顶的画面——这个想法又一次暗示了人们离开自己住所的梦想。

另一方面，这种住宿的问题根本不用考虑得那么严肃。在瑞士跨学科建筑师事务所Bureau A的设计之下，孤独的高山庇护所成了环境中不可分割的一部分，以至于看上去可以随时消失。

Bureau A的木质庇护所"安托万"以拉莫茨小说里的人物命名，这个人物曾在一块落石下生存了几周，而且，庇护所外部采用了喷射混凝土，看起来像周围山上的石头。

在海德格尔的项目案例中，庇护所充分考虑了人类的感觉体验，其外形适应当地的材料条件，同时也与庇护所内的生活风俗礼仪相匹配。但是在这里，这个庇护所期望成为悠久历史的一部分，成为当地地形以及世界构造的一部分，留下人类居住的蛛丝马迹——门、窗、烟囱，从而产生超现实的影响。

小型庇护所经常基于维持最低生活水平的目的，仅仅保证生存所需，而酒店和度假屋则以休闲为目的，或者，如之前所讨论的，越来越注重"体验"了。但是这种偏远小屋的斯巴达品质也可以应用到较为奢华的住宅建筑中。

在美国华盛顿州的山区有一座房子，由Olson Kundig建筑师事务所设计，与其之前的一些项目使用了相似的策略。选址在一处裸露的山脊上，受到严寒酷暑的侵袭。建筑师的大胆方法是设计一系列的篷子，每个都有不同的用途——公共区域、卧室、客厅和桑拿室，彼此都稍微分开摆放。

can mean the difference between life and death for those on the mountain.
The shelter has a steel structure, with a timber shell and interior, and is clad in thin concrete panels. The shelter is divided into three different units, each one with a steeply sloped roof to shed snow in winter. Each of the three units was mostly built off site and then lowered into location using a helicopter.
The dramatic setting for the shelter may be picturesque, but the image of the structure being dropped into place puts the project firmly into the tradition of the radical nomadic shelter - an image made famous by one of Buckminster Fuller's geodesic domes being transported by helicopter - and hints again at this dream of leaving one's own place.
On the other hand, this question of dwelling doesn't have to be taken all that seriously. In the hands of cross-disciplinary Swiss architects Bureau A, the lonely mountain shelter appears to become an integral part of its setting, to the extent that it seems to disappear.
Bureau A's shelter is named "Antoine", after a character in a Ramuz novel who survives weeks under a rockfall, and consists of a small timber shelter, with an outer shell of spray-concrete, designed and dressed to look like a rock of the surrounding mountains.
In Heidegger's example, the shelter incorporates deep time in a human sense, adapting its form to the local material conditions, but also the habits and rituals of life and death that go on inside. But here, the shelter wishes to pretend that it is part of a deeper time, of geology and world formation, and the effect that results from the small signs of inhabitation - doors, windows, chimney - is surreal rather than existential.

Tiny shelters frequently base themselves around the idea of an existenzminimum, of the least that is required to survive, whereas hotels and resorts operate within the realm of leisure, and increasingly, as discussed previously, "experience". But the spartan qualities of the distant hut can also be incorporated into more luxurious residential buildings.
A house in the mountains of Washington State, USA, designed

阿雷姆别墅，葡萄牙
Villa Além, Portugal

房间内部温暖奢华，但是想要在不同区域穿行，则要暴露到严酷的环境中。建筑师解释说这样设计一部分是因为客户的文化，在注重冒险的同时，肯定也重视室内的舒适度——"就像置身于睡袋中一样"，建筑师这样描述道。

但是有时候，度假屋的建筑设计会向苦行的方向发展，更注重精神层面的修行。瓦莱里奥·奥加提在葡萄牙的乡村设计并建造了一处房屋，既当作度假屋，也能在这里继续运营瑞士弗利姆斯的工作室。

奥加提的阿雷姆别墅被视为一个带有遮盖物的花园，四周由长长的混凝土高墙包围。狭长的水池位于花园的中心，两侧均种有植物。外围护结构的一端是生活区——更多的公共区域连接着花园，而一条黑黢黢的长走廊环绕生活区，可通向三间卧室，每一间卧室都有自己的小后院。

其建筑设计本身有一种出乎寻常的简朴——基础的混凝土饰面遍布整座建筑，从内至外，只有几处窗帘和小摆件来衬托生气。它看起来几乎不适合居住，但是经过许多精巧的装饰，设计感觉颇有宫殿风范，尽管也许看上去还是像住在废墟里。

奥加提将这座住宅称为"活着的景观"。在海德格尔看来，住所具有内在的传统性质，他认为房子就该是基本的形式，经过几世纪的耐心实践，住宅形式不断发展，逐渐与气候和用途呼应。奥加提心中的房屋显然与之有着完全相反的形象——一座近乎抽象的建筑，不过其愿景也是住宅的设计看上去不受时间影响。

by Olson Kundig Architects, uses similar strategies to some of the previous projects. Its site is an exposed ridge, subject to hot dry summers and harsh snowy winters. The architect's daring approach was to create a series of pavilions, one for each of the functions - public areas, the family bedrooms, guest rooms and a sauna, set slightly apart from each other.

The interior of the house is warm and luxurious, but to move between the different parts requires exposure to what can be very unpleasant conditions. The architect explains this as being part of the culture of the clients, who value adventure, and of course it accentuates the comfort of the interior spaces - "like getting into a sleeping bag" as the architect describes it.

Sometimes, however, the architecture of a retreat can move so far in an ascetic direction that it can become something almost spiritual. Valerio Olgiati has designed and built a house in rural Portugal, as a retreat from which he can continue to run his office in Flims, Switzerland.

Olgiati's Villa Além is considered as a garden with a shelter attached, and is designed as a long enclosure, held within high concrete walls. A long narrow pool is the center of the garden, with areas of planting to both sides. At one end of the enclosure are the spaces for living - more public areas are connected to the garden, while a long dark corridor loops around giving access to three bedrooms, each of which has its own smaller walled off courtyard.

The architecture itself is spartan beyond belief - a rudimentary concrete finish runs all the way through the building, inside and out, only occasionally enlivened by curtains or other items of furnishing. It looks almost uninhabitable, but through the many intellectual touches the design has something palatial about it, albeit perhaps like living in a ruin.

Olgiati describes the house as being like "landscape living". In Heidegger's vision, dwelling was an inherently traditional thing, and his vision was of a basic form that had grown through centuries of patient activity, responding both to climate and human use. Olgiati's house appears to be the exact opposite – a building of near total abstraction, yet its vision of dwelling seems equally timeless. Douglas Murphy

居住在产品中

我们为客户提供了一个建造安装都非常灵活的庇护所。这个55m²的自然庇护所设施齐全，能带人躲避城市的喧嚣。

从本意上来说，庇护所有保障基本生活的意思，它仅为功能而设计，意在满足人们"上有片瓦遮身"的原始居住需要。

VIPP庇护所的设计出发点就是返璞归真；回归自然的生活，用基本功能定义出一个紧凑的空间，其中具有典型的VIPP设计风格，即清晰一致的审美特点和对实体材料的应用。就这样，这座庇护所变成了一个便于人们逃向大自然的工具。

VIPP植根于工业产品的制造，所以"庇护所"一词成了一种住宅类型，让我们能够把这个混合物定义为宽敞的、有通用功能的可居住物体。庇护所以飞机或轮船等大体积物体为明显的参考，形成了一个宽敞的、可运输的复杂设计结构。

这个度假小屋的设计与市场上其他同类产品最大的不同在于"它并不是建筑师设计的"。与其说是建筑作品，这个庇护所更像是考虑周围环境而设计的。我们并没有从一片土地开始，根据自然环境量身定制一座房子。那附近已有不少令人惊叹的建筑了。我们想构思一个与众不同的设计：用一个完全设计好所有细节的物品带人逃离生活的压力，留给顾客的唯一选择就是决定把它放到哪里。

简单的钢网格从结构上支撑这个两层空间，只有浴室和阁楼卧室与主要的生活空间隔离开，主要生活空间包括厨房和休息区。庇护所可以同时容纳四人居住：阁楼卧室的一间主卧可以住两人，另外两人住在下面一层的沙发床上。

在这个透明的壳体内部，虽然大自然无所不在，但却不乏能提供庇护的实质掩体。这个完整的设计受大体积的飞机、轮船和潜水艇的启发，每一个螺丝都有它的用途。

当然，完美的度假屋应该地处偏远、设计巧妙却又拥有粗犷的风格，同时完全被令人惊叹的自然环境围绕。在施工中，预制化是一个基本的标准。这意味着安装过程可以在短短几天之内完成，而无需在施工地点耽搁太久。

The VIPP Shelter

Living in a Product

We have made a plug and play getaway that allows you to escape urban chaos in a 55m² all-inclusive nature retreat.

A shelter in its original sense has connotations of basic living serving a merely functional purpose and attending to our primal need of having a roof over our head.

The starting point of the VIPP shelter is going back to basics; back to nature with basic functions defining a dense, compact space wrapped in the VIPP DNA with a clear aesthetic coherence and a use of solid materials. In this way, the shelter becomes a tool that facilitates a nature escape.

VIPP庇护所

Vipp A/S

VIPP is rooted in the manufacturing of industrial objects, so the term shelter is a typology that allows us to define this hybrid as a spacious, functionally generic, livable object. Large volume objects like a plane or a ferry are a clear reference in the design, like these products the shelter is a voluminous, transportable, complex design construction.

The biggest difference between this getaway compared to anything else on the market is the fact that "I'm not an architect". The shelter is conceived more as a product becoming one with its surroundings than a piece of architecture. We did not start out with a piece of land and customised a house on it with consideration for the natural surroundings. There is plenty of amazing architecture out there. We wanted to conceive something different; an escape in the form of an object designed down to the very last detail, where the only choice left to the consumer would be where to put it.

The simple steel grid structurally supports the two level space, where only the bathroom and bed loft is shielded from the main living space comprising a kitchen and relaxing area. 4 sleeping guests can stay in the shelter at the same time: 1 main bed for 2 persons in the bed loft, and 2 additional persons on the day bed on the lower level.

Within the transparent shell, the nature is omnipresent yet with a physical blindage that provides shelter. The shelter is a finished product inspired by large volume objects such as planes, ferries and submarines, where every single screw serves a purpose.

The perfect getaway retreat should by definition be remote, smart but spartanious and surrounded by breathtaking nature. Prefabrication was an essential criteria in the construction meaning that the installation process could happen within a couple of days without an elongated process on the construction site. Morten Bo Jensen

项目名称：The VIPP Shelter
地点：Lake Immeln, Sweden
建筑师：VIPP A/S
总设计师：Morten Bo Jensen
建筑面积：55m²
外饰面：facade_primed and painted 2 mm / 0.8 in aluzink (stainless alloy) sheet metal mounted with a grid of approx. 10,000 visible screws, front edge_gutter integrated in frame, walkway_heat-galvanized and painted steel, wall and floor construction_0.6 mm / 0.02 in painted steel sheet metal with glass wool insulation and plywood shuttering, sliding glass doors_six units, panorama, skylights_remote-controlled Velux with integrated blinds, spots_10 LED spots (2 Watt) fully integrated in facade
内饰面：wall and ceiling_fitted with 3 mm / 0.1 in fire-tested wool felt covering 9 mm / 0.4 in plywood panels, bathroom ceiling_polished stainless steel, internal ridge walls by fireplace and in bathroom_powder-coated 10 x 15 mm / 0.4 x 0.6 in aluminium extrusion mounted onto a 1 mm / 0.04 in painted steel plate, floor_cast magnesite
设计时间：2013
竣工时间：2014
摄影师：©Anders Hviid (courtesy of the architect)

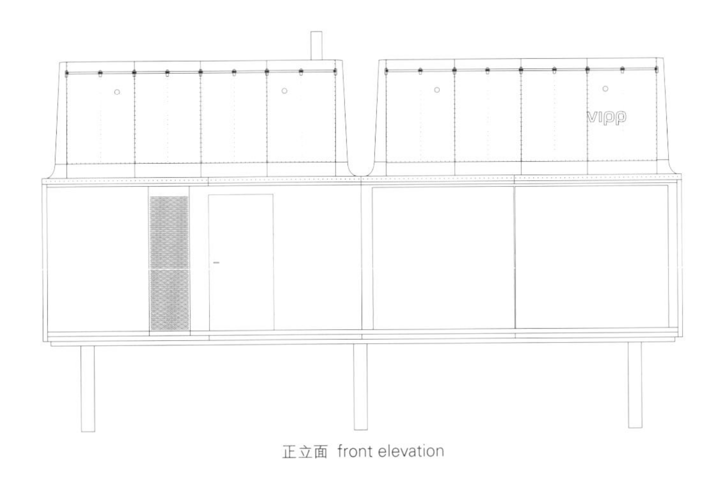

正立面 front elevation

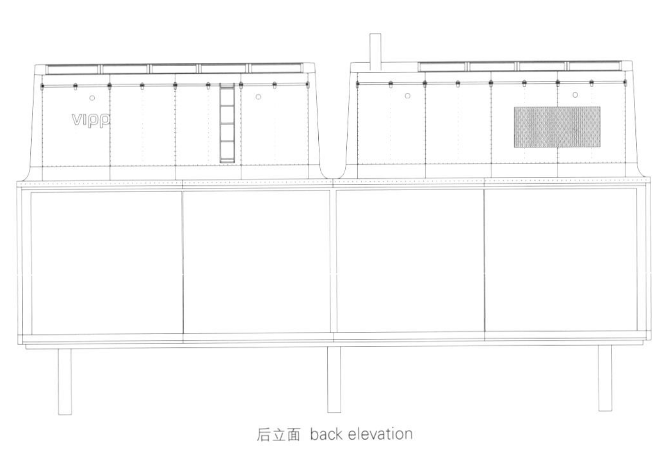

后立面 back elevation

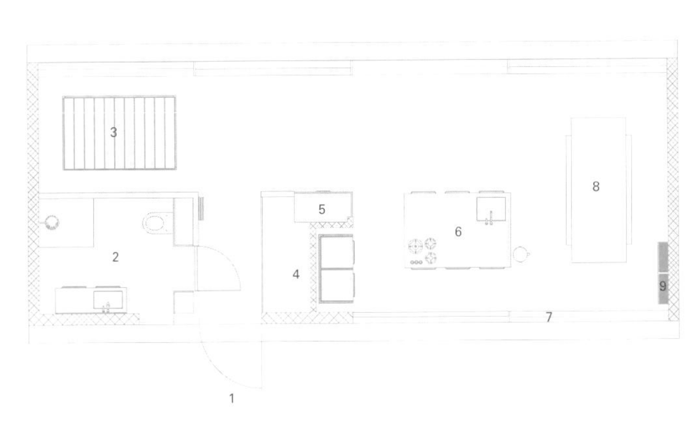

1. 入口 2. 浴室 3. 沙发床区域 4. 储藏室和洗衣房 5. 火炉 6. 厨房 7. 全景推拉窗 8. 餐厅 9. 置物架
1. entrance 2. bathroom 3. daybed area 4. storage and washing machine
5. fireplace 6. kitchen 7. sliding panorama window 8. dining area 9. shelves
一层 ground floor

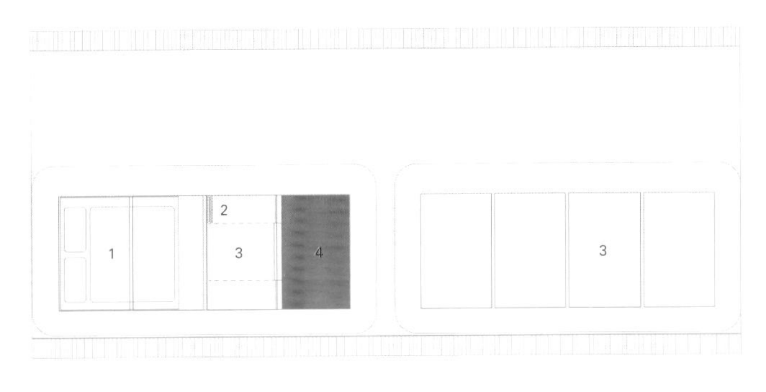

1. 阁楼卧室 2. 爬梯 3. 采光天窗 4. 储藏室
1. bed loft area 2. ladder 3. skylight 4. room for storage
二层 first floor

vipp

钓鱼小屋

Niall McLaughlin Architects

项目所在地有一个人工湖，这里原先是渔场，南部的河流是它的边界。这是一条典型的小溪流，流经汉普郡的这一部分郊区。溪流水浅湍急，格外清澈无污染，为迁徙的鳗鱼、棕鲑鱼和其他淡水鱼提供了完美的栖息地。该地区是英国境内最好的飞蝇钓区之一。

我们的客户需要一个安全的位置存放船只和钓鱼用具，这个地方同时也可以用作会议室和钓鱼者休息区。为了方便船只进出，还需要一个带顶的停泊处。在4月末到9月份的鲑鱼渔汛期，小屋就陆陆续续开始使用了。小屋四周尽可能保持开放，可以尽赏乡村自然风光。与此同时，当停止使用时，小屋也可以关闭以保证安全。

2011年，人工湖排干湖水，清理淤泥，修护湖岸。钓鱼小屋地处湖的西南角，到达河岸便捷，沿湖的风景视野也最佳。我们在1.8m深的河床中心安装了18个环形预制混凝土排水阀，在湖重新蓄水之前注满混凝土形成垫式基础。

2014年4月至9月，小屋在人工湖未重新排水的情况下，在垫式基础上进行了建设。9个镀锌钢柱框架被固定在垫式基础上。钢框架结构支撑着木地板，上层结构采用胶合橡木。屋顶采用软木椽，内部使用橡木板，外部覆盖异形铝板。

这栋建筑建有10个1.8m高的隔间。每侧的两个隔间组成了露天甲板，上方覆盖着悬挑的斜屋顶。在屋檐的百叶窗和开口接合木板构成的覆板下方，木板围合了6个中心隔间。这个外围护结构由防风雨的4个隔间和一个半封闭的储存区构成。内部的第一个隔间有一个前厅、厕所、小厨房和餐厅。另三个隔间形成了一个开放的区域，由玻璃推拉屏风围住。远处的储存区有一个存放船的阁楼、一个外部淋浴处、一个带顶的停泊处。停泊处有可移动的楼板和闸口。

在封闭的状态下，斜屋顶的形式和覆盖材料的使用直接参考了现代农业建筑的特点。外露的木结构和覆盖材料都是以橡树制成的。我们选择这种天然材料，是因为其耐用性、独特的颜色和纹理。未经处理的外部木材受侵蚀后会呈现出与屋顶覆层和钢支架一模一样的银灰色。

我们找到了一种安装在建筑侧面的百叶窗，可以在连续的水平方向上对室内产生最少的视觉侵扰。沿着屋檐向上水平旋转百叶窗，外围护结构就会消失。彼时，你将置身于水面上方的木台之上、坡屋顶之下。与外部相比，内部的木材饰面将保持温暖的金色，当周围的百叶窗都打开的时候，水中也会映出它们金色的倒影。

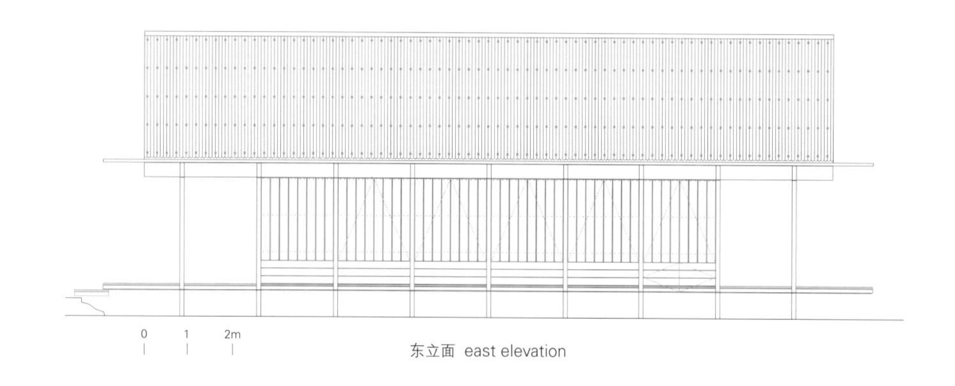

东立面 east elevation

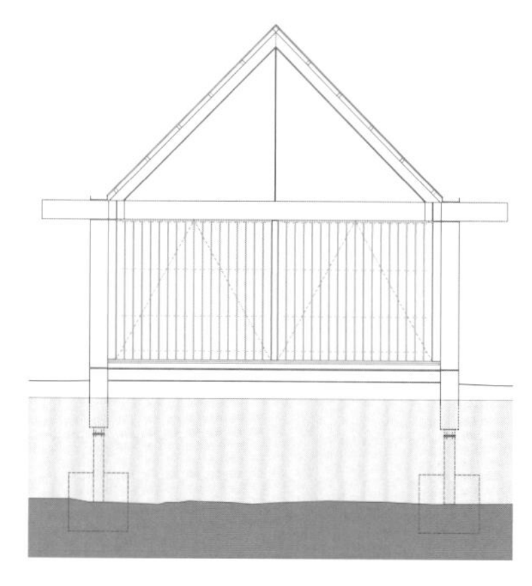
北立面 north elevation

The Fishing Hut

The site is a man-made lake, originally built as a fish farm, fed by the river that defines its southern boundary. The river is typical of the chalk streams that run through this part of the Hampshire countryside. It is shallow, fast flowing and exceptionally unpolluted, making it the perfect habitat for migrating eels, brown trout and other fresh water fish. It provides some of the best fly-fishing in the UK.

Our client wanted a secure place to store boats and fishing tackle that could also function as a meeting place and shelter for anglers. To facilitate moving boats in and out of the water a covered mooring was required. The building was to be used intermittently during the trout-fishing season from late April to September. The structure was to be as open as possible when in use to maximise views of the rural landscape in which it was situated. At the same time it had to be possible to close up and secure the building when not occupied.

In 2011 the lake was drained to carry out major works to clear silt and repair the banks. We identified a location for the Fishing Hut at the southwest corner of the lake that provided easy access from the riverbank and optimum views along the lake and river. We installed 18 precast concrete drainage rings at

项目名称：The Fishing Hut / 地点：Hampshire, United Kingdom
建筑师：Niall McLaughlin Architects
项目经理：Padstone Consulting
机电工程师：Max Fordham / 结构工程师：Price and Myers
承包商：Inwood Developments / 造价顾问：Ridge and Partners
建筑面积：66m² / 竣工时间：2014
摄影师：©Nick Kane (courtesy of the architect)

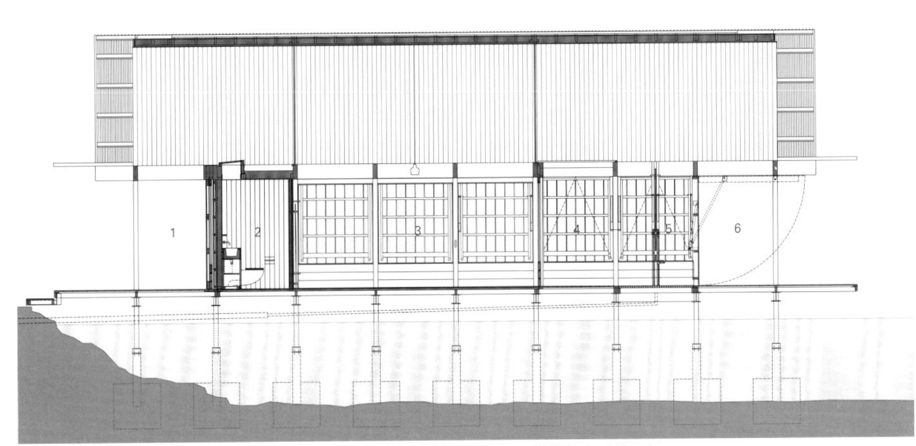

1. 入口平台 2. 浴室 3. 主要房间 4. 船库 5. 淋浴房 6. 室外平台
1. entrance deck 2. bathroom 3. main room 4. boat storage 5. shower 6. external deck
A-A' 剖面图 section A-A'

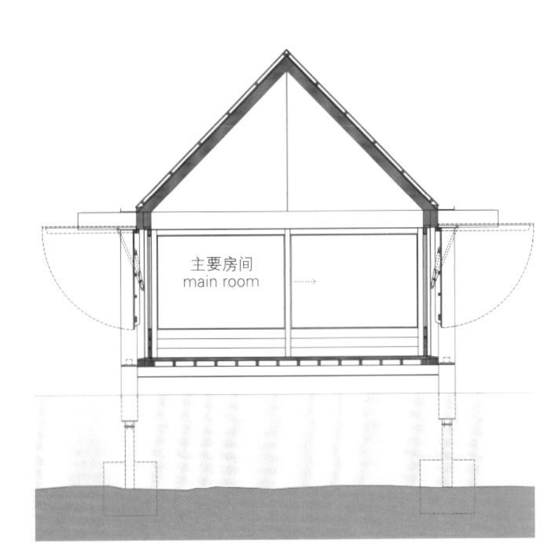

B-B' 剖面图 section B-B'

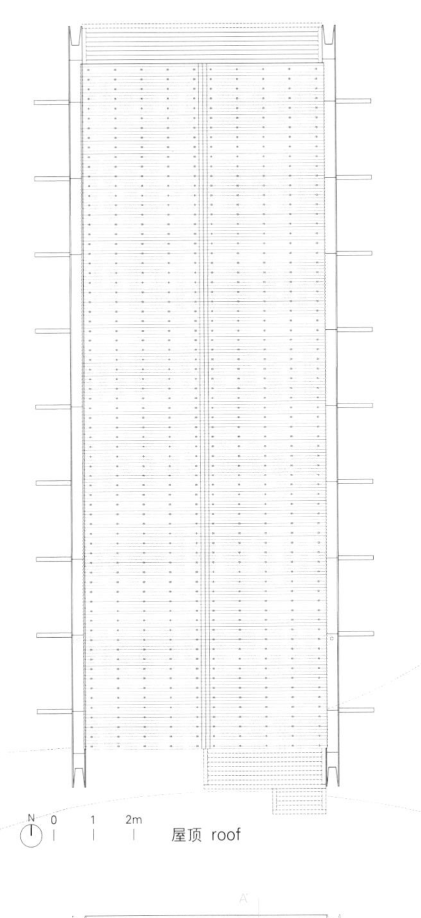

屋顶 roof

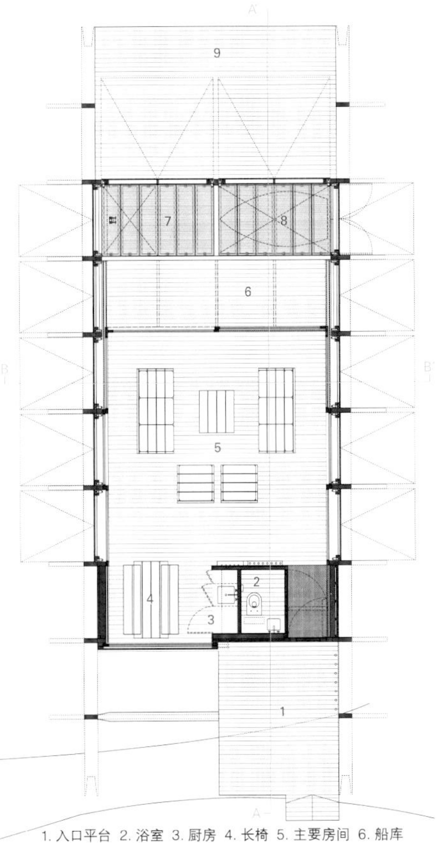

1. 入口平台 2. 浴室 3. 厨房 4. 长椅 5. 主要房间 6. 船库
7. 淋浴房 8. 小船入口 9. 室外平台
1. entrance deck 2. bathroom 3. kitchen 4. bench 5. main room
6. boat storage 7. shower 8. boat entrance 9. external deck
一层 ground floor

1.8m centers on the lakebed and filled them with concrete to form pad foundations before the lake was refilled.
The building was constructed on its pad foundations between April and September 2014 without re-draining the lake. Nine galvanized steel goalpost frames are fixed to the pad foundations. The steel frames support the timber floor structure and the glulam oak superstructure. The roof is made of softwood rafters, clad internally with oak boards and externally with profiled aluminium sheeting.
The building's structure organises its plan into ten bays of 1.8m. A pair of bays at each end form open decks, partly covered by the overhanging pitched roof. Below the eaves shutters and cladding formed of open jointed timber planks enclose the six central bays. This enclosure comprises a weather tight internal space of four bays and a semi-enclosed storage area. The first bay of the internal space contains an entrance lobby, WC, kitchenette and dining area. The other 3 bays form an open plan area enclosed by sliding glazed screens. The storage area beyond contains a loft for boat storage, an external shower and a covered mooring with a removable floor and water gate.

In its closed state the pitched roof form and handling of the cladding materials refer directly to the construction of modern agricultural buildings. The exposed timber structure and cladding is made of oak. We chose this native species for its durability and characteristic colour and grain. The untreated exterior timber will weather to match the silver-grey colouring of the roof cladding and steel supports.
Along the sides of the building we sought a form of shutter that gave the least visual intrusion on the continuous horizon. By horizontally pivoting the shutters upward from the eaves the enclosure disappears, leaving you on a deck above the water beneath the pitched roof. In contrast to the exterior, the timber of the enclosed interior will retain in its warm golden tone, which is revealed and reflected in the water as the perimeter shutters are opened. Niall McLaughlin Architects

斯古塔高山庇护所

OFIS Arhitekti

来自OFIS建筑师事务所的Rok Oman和Spela Videcnik共同领导哈佛大学设计研究生院一建筑工作室完成了这个项目。2014年秋天，这个由13名学生组成的工作室面临挑战，需要在高山地区的极端气候条件下设计出一个新型但实用的居住空间。受丰富多样的斯洛文尼亚本土建筑遗产的启发，学生们提出了12个满足各种场地条件、材料条件以及设计要求的方案。

山区的极端气候条件为建筑师、工程师和设计师的设计提出了挑战。在容易遭受极端环境因素影响的环境下，建筑物能经得起极端天气、剧烈的温差变化和崎岖的地形所产生的影响都是非常重要的。与环境协调一致的设计不仅是一种保护措施，也直接关系到切身安全问题。风、雪、山体滑坡、地形、气候等恶劣条件需要特定的建筑形式和概念设计来应对它们。这个新项目取代了先前一个在这里建了50年的露营地。

这个营地庇护所代表了人类的基本需求。它是一个避难所的象征。外部形态的设计以及材料的选择都是为了应对极端的山区条件，并要有开阔的视野。其选址位于荒野中，所以需尊重自然资源，因此建筑师采取了刚柔并济的态度，既要让庇护所牢牢固定在地面上，又要确保对地面产生最小的影响。此外，外壳必须使用非常耐磨的材料，再与里德的薄玻璃纤维öko混凝土表层构件搭配，可满足所有美学的要求、材料质量的要求，并能够承受特别严酷的气候条件。室内设计要求端庄得体，能够容纳多达八名登山者并为他们提供庇护。本设计由三个模块组成，既方便运输，又能根据功能划分空间。首先是入口、储藏处及准备食物的小空间。第二个是提供睡觉和社交的空间。而第三个是双层睡眠区。从两端的窗口可欣赏到山谷和斯古塔山的壮观美景。

由于其独特的安装过程，庇护所被设计为一系列模块，那样就可以一部分一部分地把它们运到山上了。整个原型是在车间建造的。模块被设计为一系列结实的框架，可以去现场组合起来，其安装简便并且可以减少地基对土地的不良影响。为尽可能不影响山区环境，这些模块用销钉连接，钉子的位置都是精心设计过的，同时以此作为营地的地基。玻璃为三层玻璃系统，这是经过计算的，可承受预计的强风和雪荷载。该营地的安装是在Matevz Jerman的指导下由PD Ljubljana Matica执行的，由斯洛文尼亚武装部队和卢布尔雅那服务站的一支山地救援队承担直升机运输任务。整个运输和安装过程只花了一天。

Alpine Shelter Skuta

The project was developed from an architectural design studio at the Harvard Graduate School of Design led by Rok Oman and Spela Videcnik from OFIS Arhitekti. In fall 2014, the studio of thirteen students were facing the challenges of designing an innovative yet practical shelter to meet the needs of the extreme alpine climate. Inspired by the vernacular architecture of Slovenia with its rich and diverse architectural heritage, the students produced twelve proposals meeting various site conditions, material considerations, and programmatic concerns were produced and cataloged.

The harsh climatic conditions in the mountains introduce a design challenge for architects, engineers and designers. Within a context of extreme risk to environmental forces, it is important to design buildings that can withstand severe weather, radical temperature shifts, and rugged terrain. Responding to environmental conditions is not only a protective measure, but also translates into a matter of immediate life safety. The harsh conditions of wind, snow, landslides, terrain, and weather require a response of specific architectural forms and conceptual designs. The new shelter is replacing a 50 year old bivouac that had previously been on the site.

The bivouac shelter represents a basic human necessity. It is a symbol of refuge. The outer form and choice of materials were chosen to respond the extreme mountain conditions, and also provide views to the greater landscape. Its position within the wilderness requires respect for natural resources, therefore must meet the ground in a light and firm manner

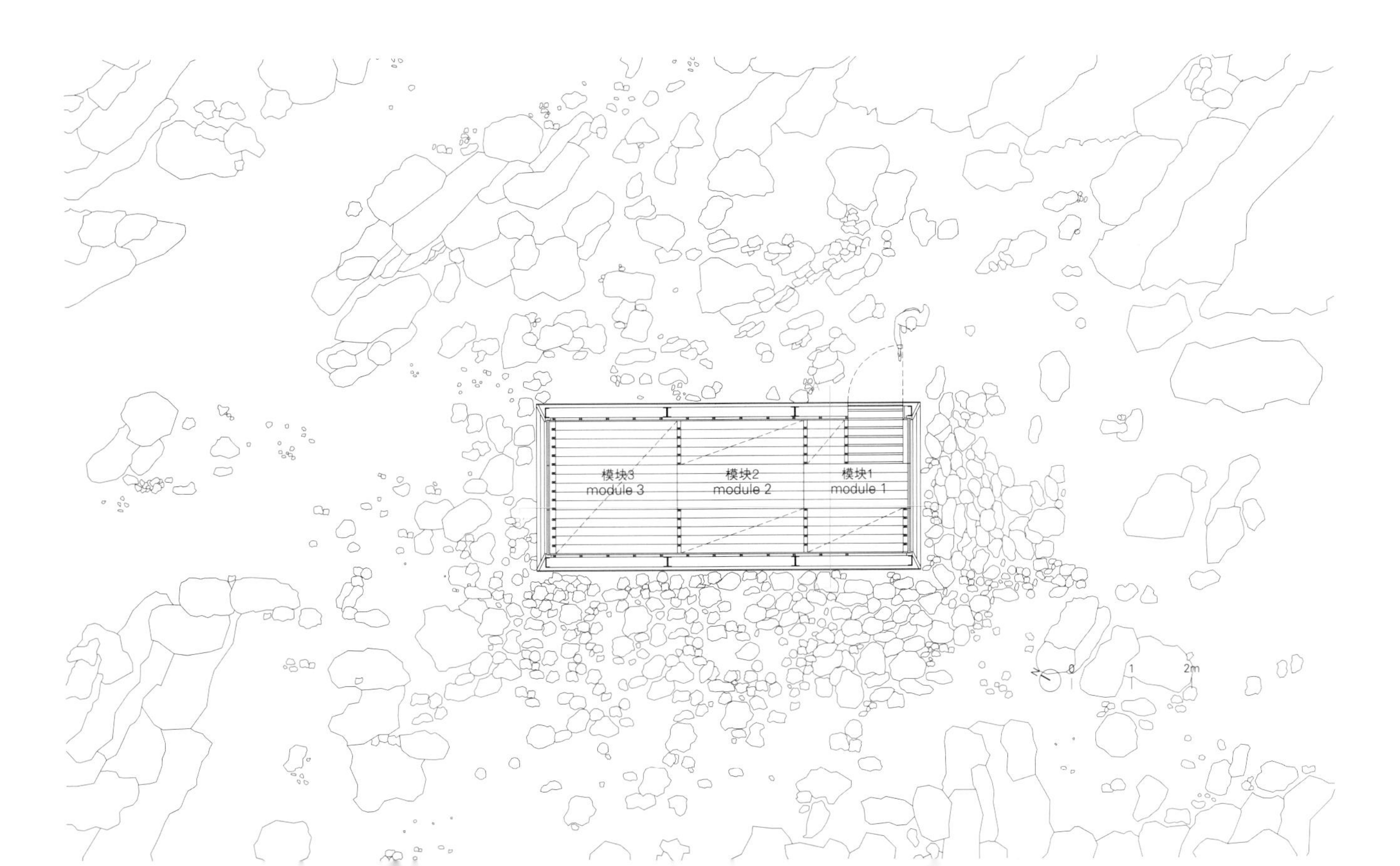

南立面 south elevation

东立面 east elevation

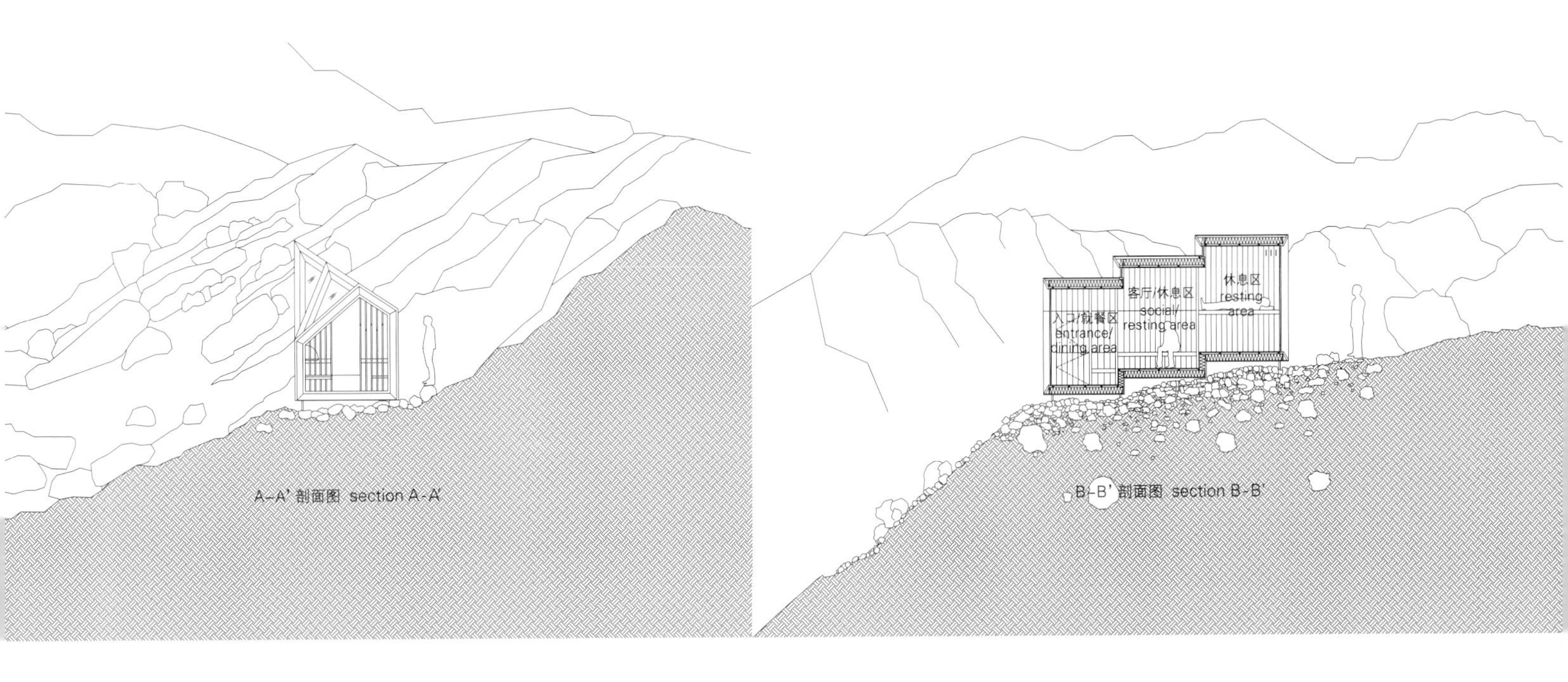
入口/就餐区
entrance/
dining area
客厅/休息区
social/
resting area
休息区
resting
area
A–A' 剖面图 section A-A'
B-B' 剖面图 section B-B'

to ensure the shelter is strongly anchored while having a minimal impact on the ground. In addition, the outer shell needed to be realised with a highly resistant material and in collaboration with Rieder thin glass fibre öko skin concrete elements were able to meet all the requirements of aesthetics, material quality and the challenge of being able to withstand especially rigorous weather conditions. The design of the interior dictates modesty, totally subordinate to the function of the shelter providing accommodation for up to eight mountaineers. The design consists of three modules, in part to allow for transport and also to programmatically divide the space. The first is dedicated to the entrance, storage and a small space for the preparation of food. The second one provides space for both, sleeping and socializing while the third features a bunk sleeping area. Windows at both ends offer beautiful panoramic views of the valley and Skuta Mountain.
Due to the nature of the installation process, the shelter was designed as a series of modules so that it could be brought to the mountain in parts. The entire prototype was constructed off-site in the workshop. The modules were planned as a series of robust frames, which would then be braced together onsite providing a manageable installation and a less invasive foundation. In order to keep the mountain site as undisturbed as possible, the modules are fastened onto strategically placed pin connections, which also act as the foundation on the site. The glass is a triple pane system that has been calculated to withstand to the projected strong wind and snow loads. Installation of the bivouac was carried out by PD Ljubljana Matica under the direction of Matevz Jerman, helicopter transport while the Slovenian Armed Forces and a team of Mountain Rescue Service - station Ljubljana. The whole transportation and installation process was carried out in one day.

项目名称：Alpine Shelter Skuta
地点：Mountain Skuta, Slovenia
建筑师：Rok Oman, Spela Videcnik_OFIS Architects
合作方：AKT II, students at Harvard University Graduate School of Design, Freeapproved and PD Ljubljana Matica
哈佛大学设计研究生院的学生设计团队：Frederick Kim, Katie MacDonald, Erin Pellegrino
OFIS建筑师团队：Andrej Gregoric, Janez Martincic, Maria Della Mea, Vincenzo Roma, Andrea Capretti, Jade Manbodh, Sam Eadington
结构工程：AKT II (Hanif Kara, Edward Wilkes)
当地结构工程师：Projecta, Milan Sorc
工程与咨询：Freeapproved, Anze Cokl
立面（öko表皮）：Rieder Smart Elements (Wolfgang Rieder, Matthias Kleibel)
玻璃：Domen Komac_Guardian
客户：PD Ljubljana Matica / 用途：shelter / 竣工时间：2016
摄影师：
©Andrej Gregoric (courtesy of the architect) - p.40
©Anze Cokl (courtesy of the architect) - p.38~39, p.45, p.46, p.48~49
©Janez Martincic (courtesy of the architect) - p.42~43, p.44

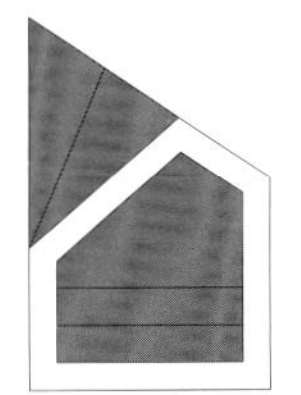
模块1 module 1

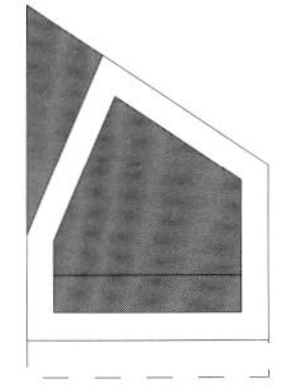
模块2 module 2

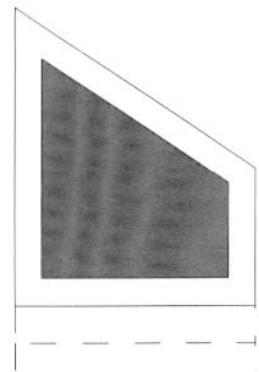
模块3 module 3

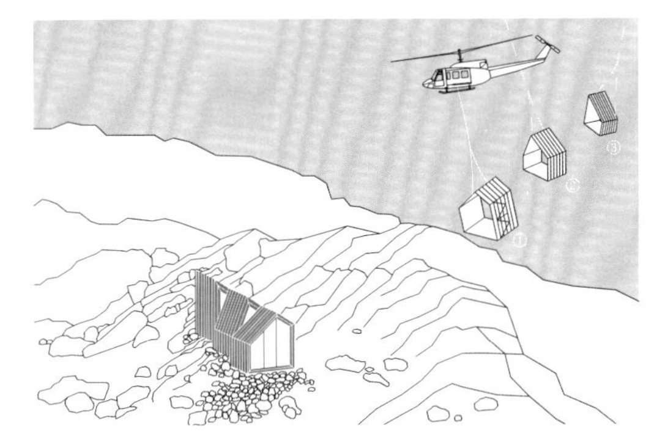

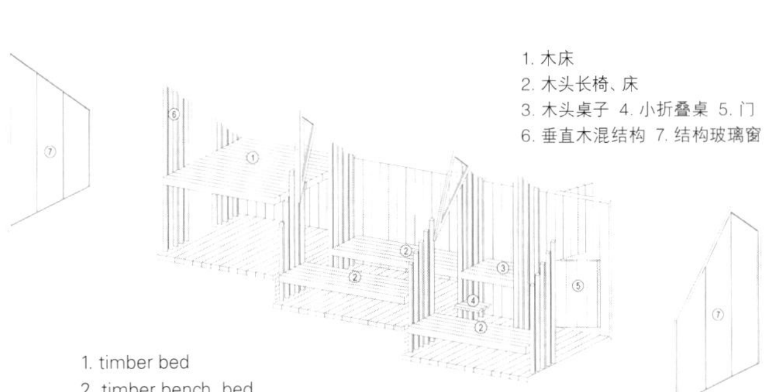

1. 木床
2. 木头长椅、床
3. 木头桌子 4. 小折叠桌 5. 门
6. 垂直木混结构 7. 结构玻璃窗

1. timber bed
2. timber bench, bed
3. timber table 4. small folding table 5. door
6. vertical timber beton 7. structural glass window

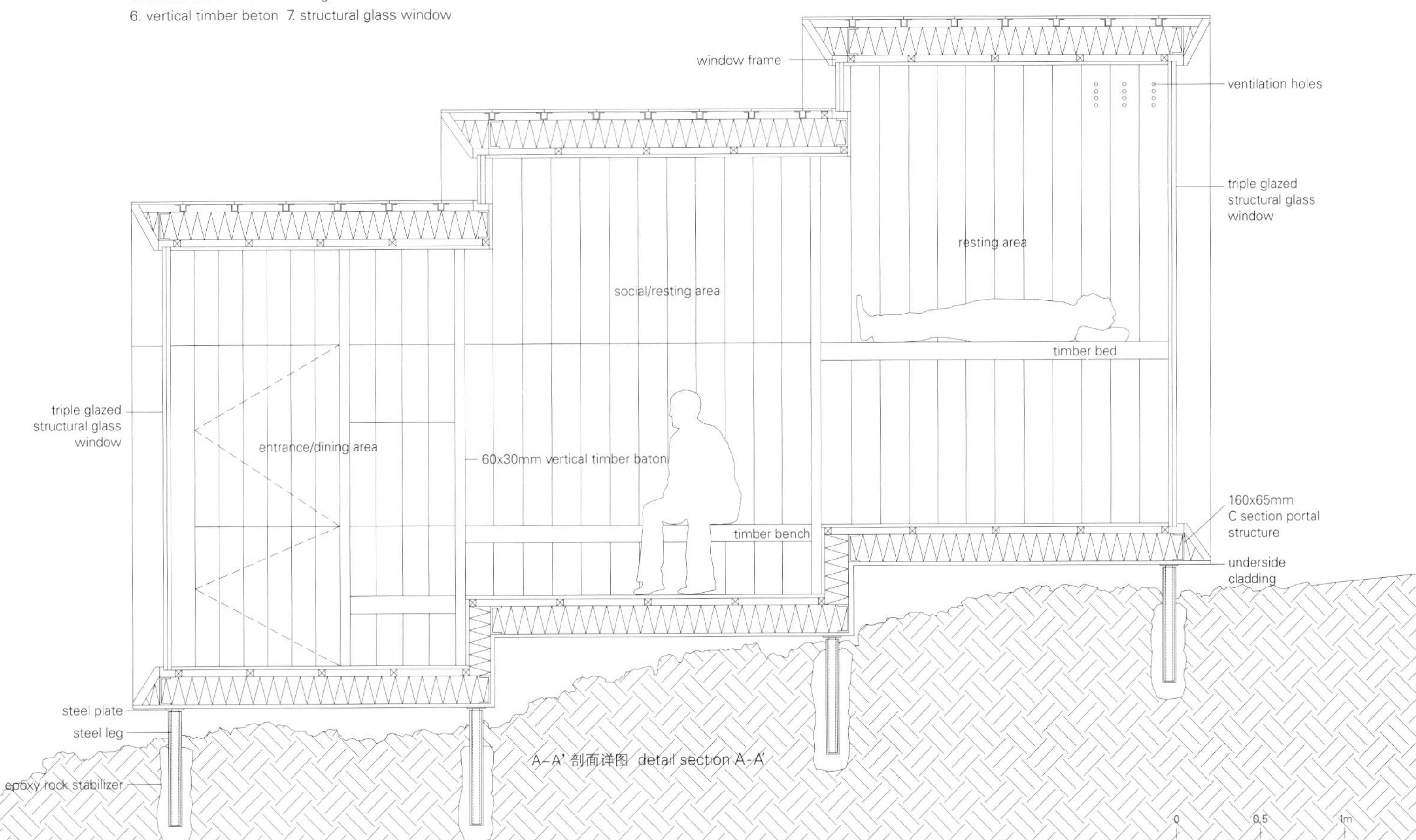

A-A' 剖面详图 detail section A-A'

这座山既令人着迷又令人畏惧。长久以来，瑞士有一种根深蒂固的传统，那就是观察阿尔卑斯山，与它们共存，并隐匿于它的身后。这个不朽的风景令人敬畏，又让人挂念，它的吸引力在瑞士著名作家查尔斯·斐迪南·拉莫茨的著作中有所体现。他的小说《德布朗斯》中描述了1714年里齐尔尼山谷发生了一次大规模的岩石崩落事故，岩石堆满了山谷牧场。小说的主人公安托万在逃生成功之前在岩石下面生存了七周。

"安托万"庇护所对于作者和这次阿尔卑斯之行来说是一种悼念。这个小木屋足够容纳一个成人居住，它隐藏在一块突出的岩石里面。该项目借鉴了瑞士长久以来的这种隐藏式地堡的传统，也结合了阿尔卑斯地区最具有城市化气息的风景。早在1975年，法国哲学家保罗·维希留就描述过，这种按照隐藏原则建造的军事建筑已经吸引建筑师很久了。

"安托万"庇护所项目营造了一种不太稳定的高山庇护所（只保障最低生存权），使用时很容易被破坏，人们可以随意进入并藏身其中。它拥有基本的建筑元素——壁炉、床、桌子、凳子、窗户——但是来访者也要承担一些风险，因为这个岩石状的庇护所就坐落在当年的岩崩地带。

"安托万"庇护所是由韦比耶3D基金会委托设计的艺术家驻地。一开始建在小村落里，后来移到海拔较高的雕塑公园。6周的居住时间促成了这个庇护所的建造，这里可以被当成一个居住地，遵循了安德烈·布洛克所描述的建筑雕塑传统，并且由安德烈·布洛克和克劳德·帕朗对其进行实地开发。

Antoine

A mountain has a power to call for feelings of fascination and fear at the same time. Switzerland has a strong tradition of observing the Alps, living with them, hiding inside them. The awe and the anxiety that this monumental landscape appeals is reflected in the writings of Charles-Ferdinand Ramuz, one of the most important Swiss writers. His novels, Derborence, describes the massive rock fall that covered the pastures of the valley of Lizerne in 1714. Antoine, the main character, survives

"安托万"庇护所

Bureau A

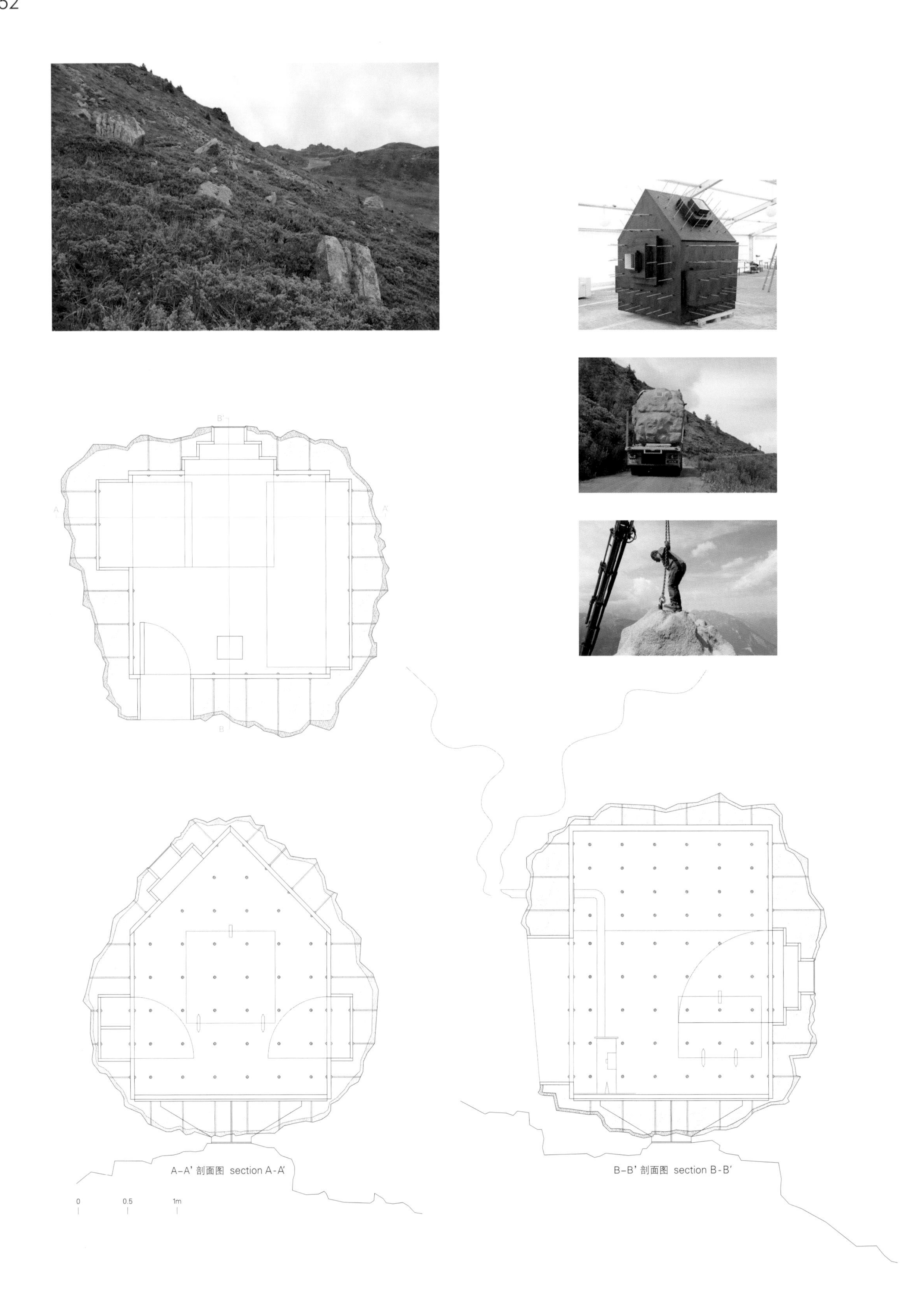

A-A' 剖面图 section A-A'

B-B' 剖面图 section B-B'

项目名称：Antoine
地点：Les Ruinettes, Verbier, Switzerland
建筑师：Bureau A
委托方：3D Foundation Verbier
建筑面积：4m²
设计时间：2014
施工与竣工时间：2014
摄影师：©Dylan Perrenoud (courtesy of the architect)

seven weeks under the rocks before he manages to reach his village, and life.

Antoine is a tribute to the alpine experience and to the writer. The small wooden cabin, big enough for the life of one man, is hidden inside a projected concrete rock. Referring to the long lasting Swiss tradition of hidden bunkers, the project integrates the highly urbanised landscape of the Alps. Already described by the French philosopher Paul Virilio in 1975, mili-

tary architecture conducted by principles of camouflage has, for long, fascinated the architects.
Antoine creates an alpine shelter, a precarious "Existenzminimum" somewhat subversive in its use where one can freely enter and hide. It contains the very basic architectural elements – fire place, bed, table, stool, window – but demands to the visitor some risk taking as the rock hangs literally on the rock fall field.

Antoine was a commission by the artist residency Verbier 3D Foundation. It was self-built in the village and transported to the high-altitude sculpture park. The 6 weeks residency allowed for the construction of what can be considered an inhabited sculpture that follows the tradition of architecture-sculpture described by artist André Bloc and developed physically by André Bloc and Claude Parent.

卡尔德拉住宅

DUST

这座自给自足的独立住宅坐落在南亚马逊圣拉斐尔流域卡内罗山的西南山麓冲积平原上，这里杳无人烟，距亚利桑那州的图森东南部有两小时车程，距美国和墨西哥边界北部24km。它坐落在广阔的草原和面向西方的远山之间，远离了那些可能会冒险穿越这片风景的人。由于靠近边界，这里会有不少移民经过，所以业主要求设计了这种无法侵入的结构。

房子从天然的草地、金刚砂橡树和广阔的草原中浮现，呈简单的矩形，45.7cm厚的墙壁由岩浆和水泥浇筑而成。建筑材料由粉碎的轻质红色矿渣、水泥和水混合而成，然后在模板中将其夯实。这些墙壁构成了这个结构，并且集隔热和保温功能于一体。

这个87.8m²的建筑灵感来源于当地一种被称为"门廊"的传统房屋结构。在平面设计中将两个卧室与客厅相对，中间是门廊。门廊两端是一扇大的折叠门，与外面相连，可以打开以保证外面的自然光线透进来，也可以关闭以保证安全。

门廊和窗户洞口提供了天然通风，能使屋内凉爽起来，而薪柴燃烧又能提供热量。水是从井里打出来的，小型家用电器通过太阳能供电。整个建筑完工后清理掉了约22.9m³的废料。

Caldera House

The off-grid house is located in a remote landscape on the southwestern bajada of the Canelo Hills in Southern Arizona's San Rafael Valley, two hours southeast of Tucson and 15 miles north of the US and Mexico Border. Siting is balanced between the prospect of the open range and distant mountains toward the west and refuge from those who may venture across the landscape. Proximity to the border and immigrant related foot traffic led the owner to request an impenetrable structure.

The house emerges from the native grasses, Emery Oaks,

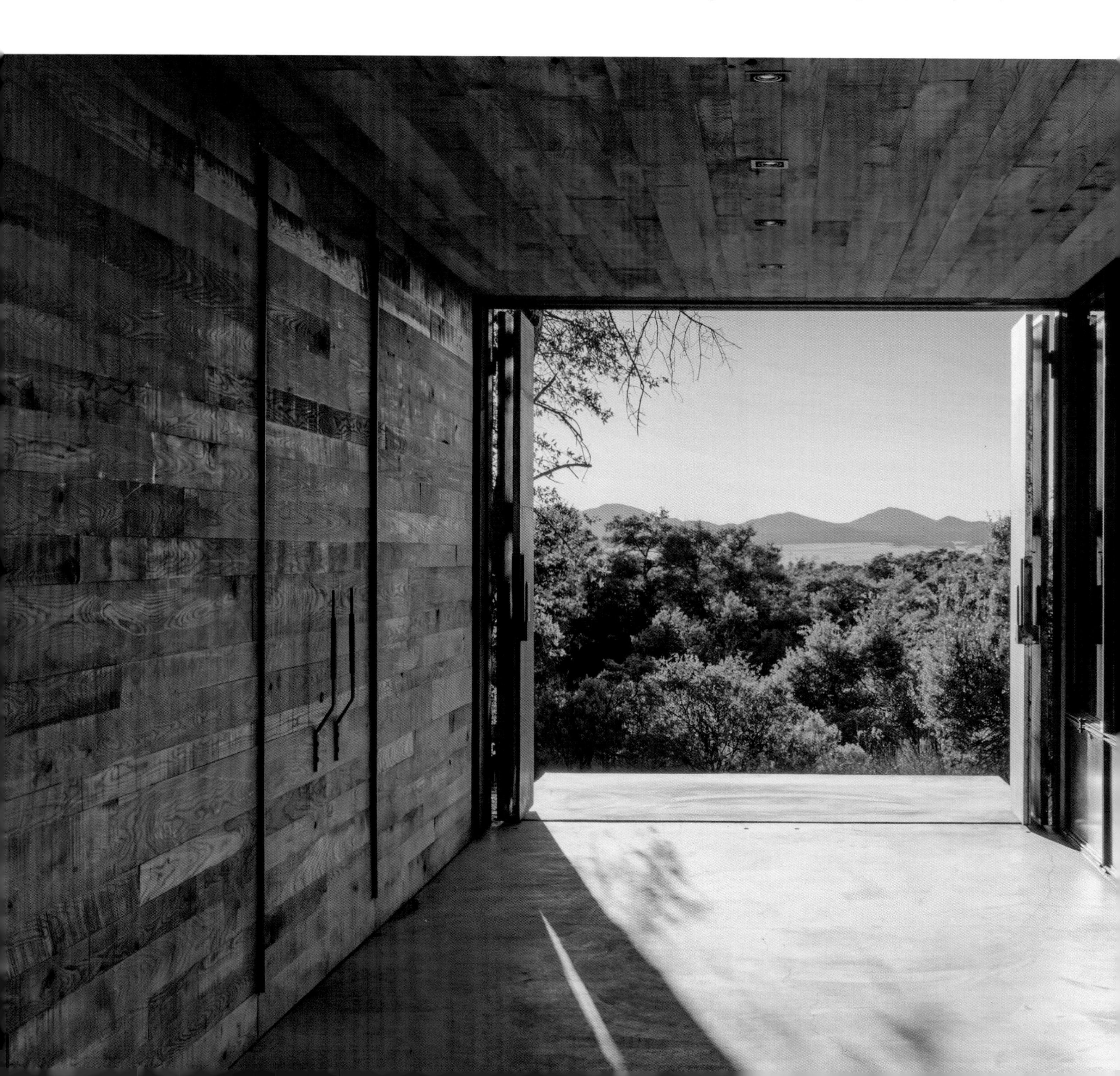

and open ranges beyond in a simple rectangular form of 18" mass walls constructed of poured lava-crete. The material is comprised of a mixture of pulverized lightweight red scoria, cement, and water, rammed into formwork. These walls create the structure, finish and offer insulation and thermal mass all in one stroke.

The 945 sq. ft. structure takes clues from a vernacular "zaguan" housing typology. The plan locates two bedrooms opposite a living room; a zaguan runs between them. Large bi-fold doors on the ends of the zaguan connect the space to the outside, introducing natural light when open, and security when closed.

Cooling is provided by natural cross ventilation through the zaguan and window openings, while wood fuel sourced on the property provides heating. Water is from a well, while solar power is used for minimal electrical and appliance needs. A single 30-yard rolloff of waste was removed after the entire construction process.

项目名称：Caldera House
地点：Southeast Arizona, San Rafael Valley
建筑师：DUST
总建筑师：Cade Hayes, Jesus Robles
项目团队：Jay Ritchey, Agustin Valdez, Ben Gallegos
顾问：Paul Schwam of Lava Works Concrete
屋顶：Flashings INC.
总承包商：Jesus Robles, Cade Hayes
建筑面积：98.5m²
室内面积：49m²
竣工时间：2015.12
摄影师：©Jeff Goldberg/Esto (courtesy of the architect)

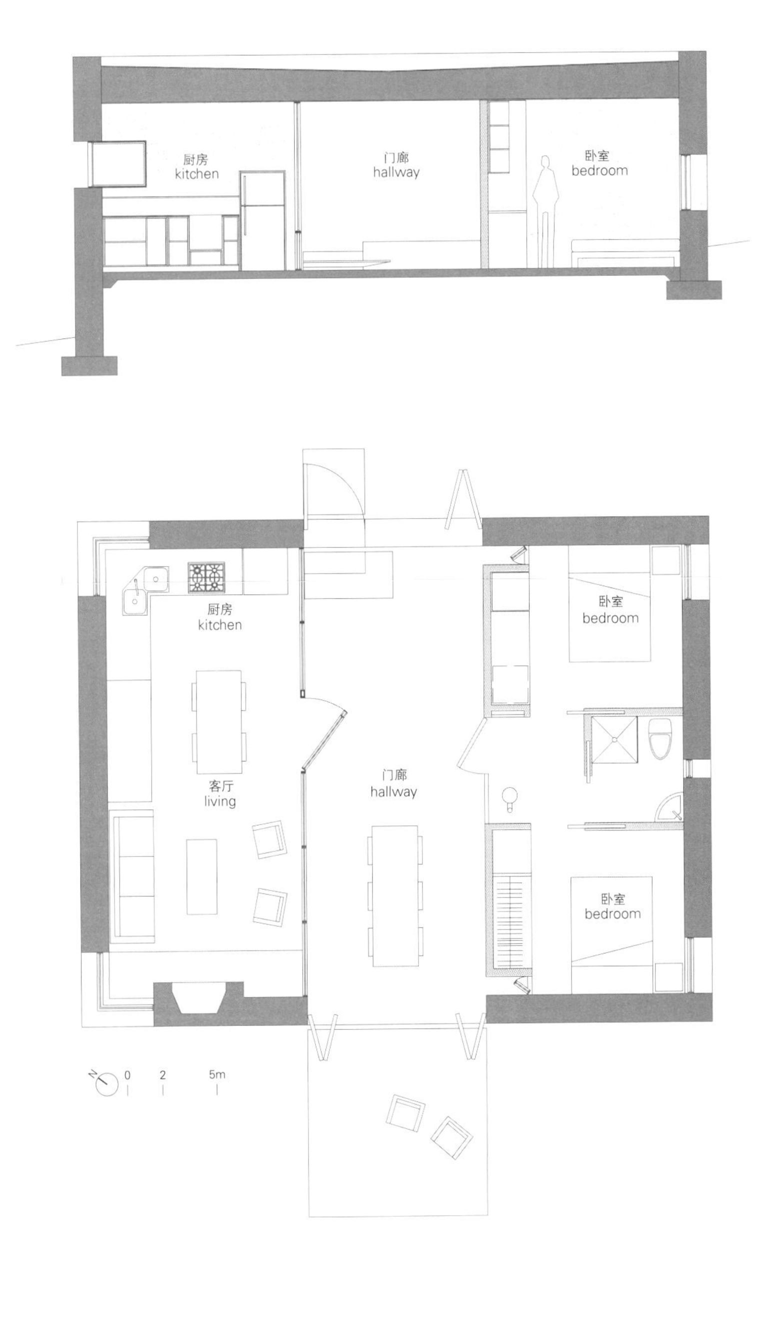

阿雷姆别墅

Valerio Olgiati

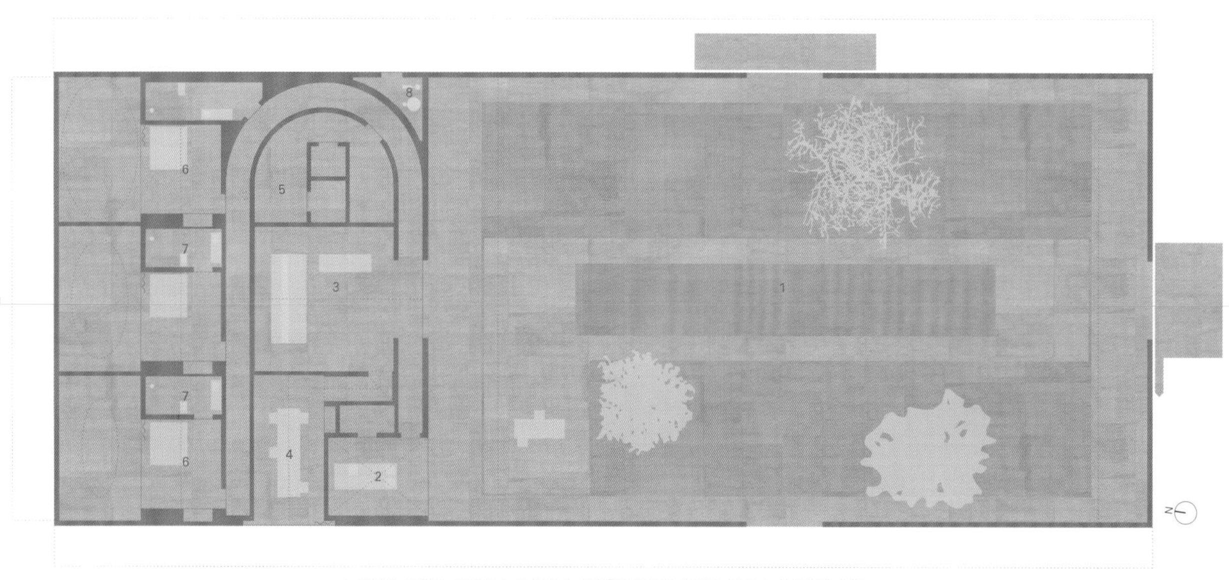

1. 庭院 2. 厨房 3. 起居室 4. 办公室 5. 房屋设备空间 6. 卧室 7. 浴室 8. 泳池设备空间
1. courtyard 2. kitchen 3. living room 4. office 5. house technical space 6. bedroom 7. bathroom 8. pool technical space

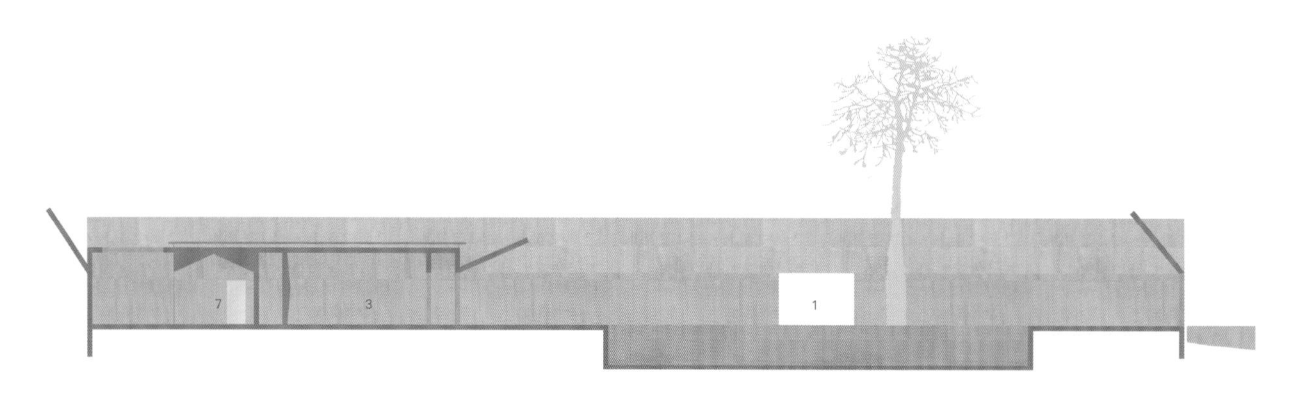

这个项目位于阿连特茹，大西洋内陆10km处。这里以丘陵地带和田园风光为特色，并且遍地栽种着美丽的栓皮栎。

这里的气候温和干燥。设计的主要意图是创造一个与世隔绝的花园。它的围墙高达5.5m，可以乘凉，它给人的整体印象是沙漠中的一员，干燥、多石而落满灰尘。项目全部由微红的现浇混凝土建造而成。

这些围墙很好地诠释了这个建筑群的特点，它给人留下的印象是向天空开闭的花瓣。住所本身是十分隐蔽的，在围墙背后延展开来。

客厅位于南北中轴线的一端。在这里能眺望池塘，并且可以通过花园南边的大门看到平坦而空旷的风景。弯曲的走廊让居住者能够退回到暗处，走进私密的房间。

Villa Além

This project is located in Alentejo about 10 km inland from the Atlantic Ocean. The area features a hilly, rural landscape and is covered with beautiful old cork oaks.

The climate is mild and dry. The primary intention here is to create a secluded garden. The surrounding walls are up to five and a half meters high to provide the necessary shade and the entire impression created is one of a desert, dry, stony and dusty. Everything is constructed from slightly reddish, in situ concrete.

The character of the complex is chiefly defined by the walls, which create the impression of petals that close and open towards the sky. The dwelling itself is invisible and develops across a single floor behind the surrounding walls.

The living room is located at the end of a strict axis leading from north to south. It overlooks the pool and offers a view through the southern door in the garden wall across a flat and empty landscape. A curved hallway allows the inhabitants to retreat into shadows and into the introverted private rooms.

项目名称：Villa Além / 地点：Alentejo, Portugal / 建筑师：Valerio Olgiati / 合作方：Patricia Da Silva_Project Manager of Office Olgiati, Daisuke Kokufuda, Liviu Vasiu
总承包商：Matriz Sociedade de Construções Lda / 材料：in-situ concrete, steel, marble / 用途：house / 设计时间：2009 / 施工时间：2013.1 / 竣工时间：2014.4
摄影师：©Archive Olgiati

种马住宅

Olson Kundig Architects

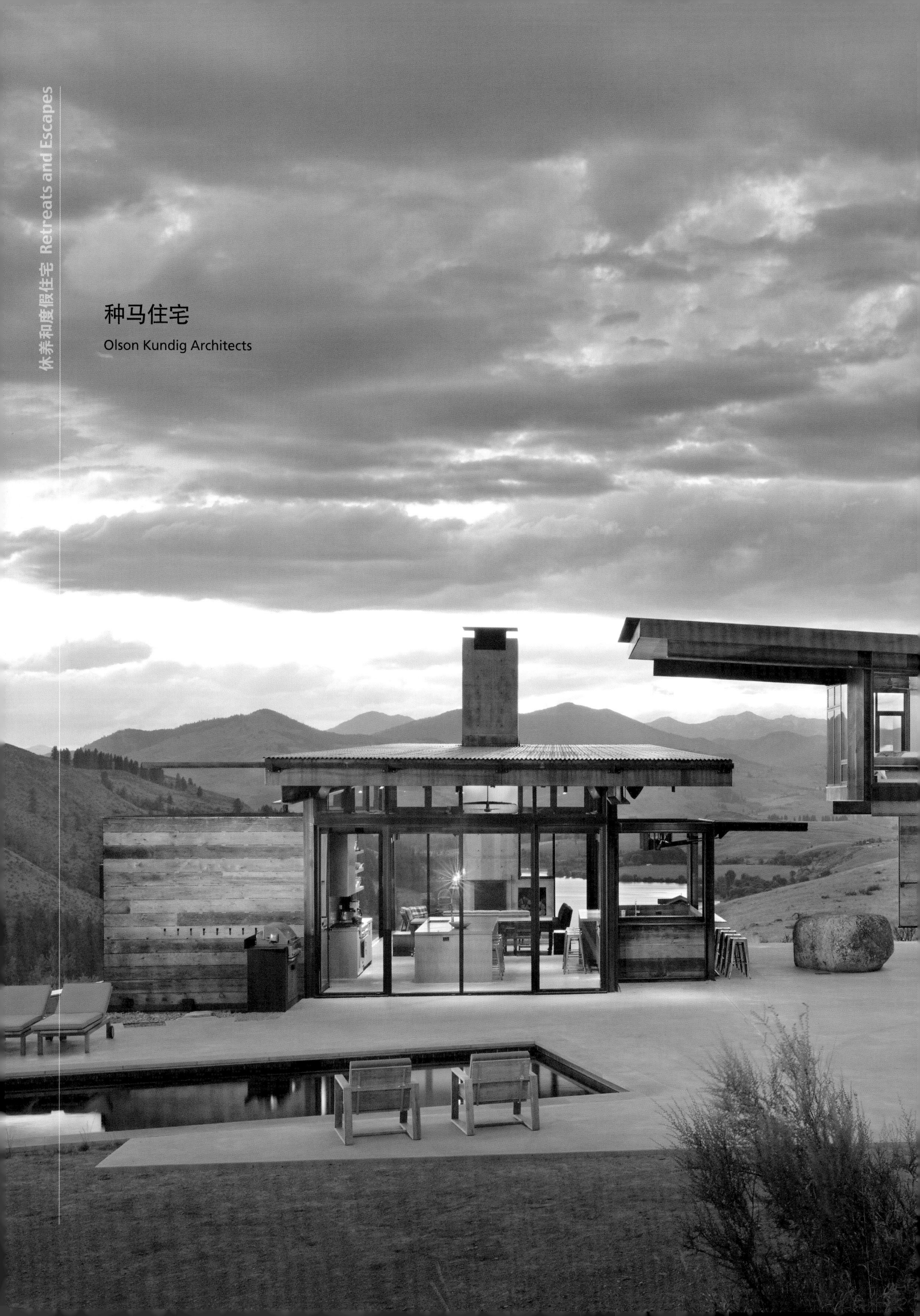

这个96.6km长的梅索谷冰川位于华盛顿州的瀑布大区北部，是一处奇特的景观，这里夏季炎热，特别容易引发火灾，冬季会有厚重的积雪。这里是一个真正四季分明的景点，我就在这里长大。我们的客户想在这里建造自己的第二个家，对他的家人和孩子来说有一点冒险风格的那种家。他们是一对伟大的父母，经常有意识地把冒险经历当作家庭的回忆。

在这里，与冒险有关的房子就是要逼你出门并且积极应对各种季节。你必须走出去融入它。所以，这所房子有一定的不便之处，但是我和我的客户都把这些看作美妙而难忘的瞬间。冒险本身就是不便的，它会强调并提醒你身在何处。我经常爬山，这种经历似乎很浪漫，但也会让你感到不适，甚至是恐怖。你会冷、会热、会痛。但是为什么人们在经过理智的思考之后还会去做呢？因为它会让你对生活充满激情。

这所住宅由四座建筑物组成，以中央的庭院和池塘为中心，它们并没有连接在一起，但它们组合在一起像环绕的马车。住宅就像一个小型的野营地，你需要穿过一个个帐篷。外部材料十分坚实，因为要适应沙漠化的气候。但内部却很温馨舒适，就像进了一个睡袋一样，温暖干燥而有安全感。

每座建筑都拥有令人惊叹的视野，能看到周围的种马山脉和Pearrygin湖。家庭活动室、厨房、酒吧等"公共"区域都组合在一起，这个结构几乎是全开放式的，周围是可滑动的玻璃墙。主卧、儿童卧室和密室都在另一座建筑里。第三座建筑是客人的房间，这使得客人可以有自己的隐私空间。桑拿房位于第四座建筑，这里不但私密性强，还能欣赏到独特的风景，可以眺望山谷。

项目刚开始的时候这里还是空空荡荡的，让我们拭目以待吧！这是一块冰川漂砾（冰川退去后留下的岩石），这也成了整个计划的核心点。它的位置相对靠近酒吧，因此也成了可以坐着或者放酒的地方。我把它视为一个大件家具。设计包含了家庭活动和娱乐。例如，电视墙可以朝里，也可以向外朝向庭院。如果电视在播放一场体育比赛的话，每个人都可以在泳池里看到它。酒吧附近还有一个甜筒岛游乐场，只需按一下按钮墙就能打开。这成了这次建筑设计中尤为成功的一部分。大家会自然而然地在酒吧附近闲逛，这里的墙能开启，对于我们想要实现的内外关系来说是很重要的。

整个项目使用的木质护墙板是从斯波坎市的一个旧谷仓捡回来的。每块木头不同的颜色都反映了它的历史及其用处。玻璃是非常重要的。我从不喜欢待在与外界毫无联系的屋子里。我喜欢在符合建筑规范的情况下把玻璃聚在一起使用，使其与自然风景生动地联系在一起，而不是像辣椒面那样撒得到处都是。

项目名称：Studhorse / 地点：Winthrop, Washington, USA
建筑师：Olson Kundig Architects
项目团队：Tom Kundig (FAIA)_Design Principal, Mark Olthoff (AIA, LEED AP)_Project Manager, Gus Lynch (LEED AP)_Staff, Debbie Kennedy (LEED AP ID+C)_Interiors
结构工程师：MCE Structural Consultants
钢结构建筑师：Argent Fabrication
混凝土工程师：Westlake Concrete / 钢结构工程师：Alpine Welding
室内设计：Olson Kundig / 承包商：Schuchart/Dow
建筑面积：4,078m² / 竣工时间：2012
摄影师：©Benjamin Benschneider (courtesy of the architect)

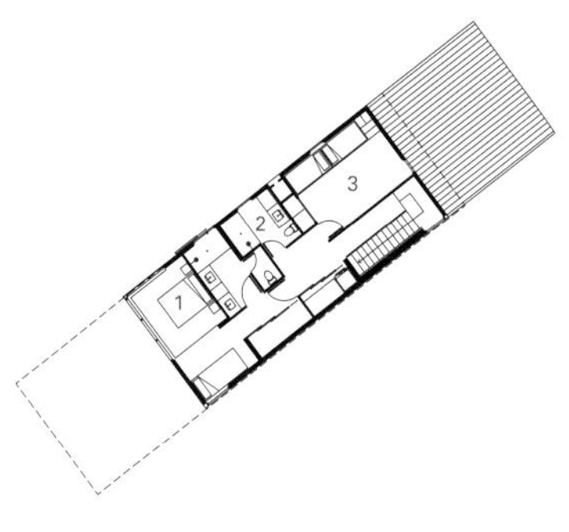

二层 first floor

1. 主卧
2. 浴室
3. 卧室
4. 沾泥物品寄存室
5. 客房
6. 厨房/小酒吧
7. 起居室/餐厅
8. 密室
9. 泳池/热水浴缸

1. master bedroom
2. bath
3. bedroom
4. mudroom
5. guest room
6. kitchen/wet bar
7. living/dining
8. den
9. pool/hot tub

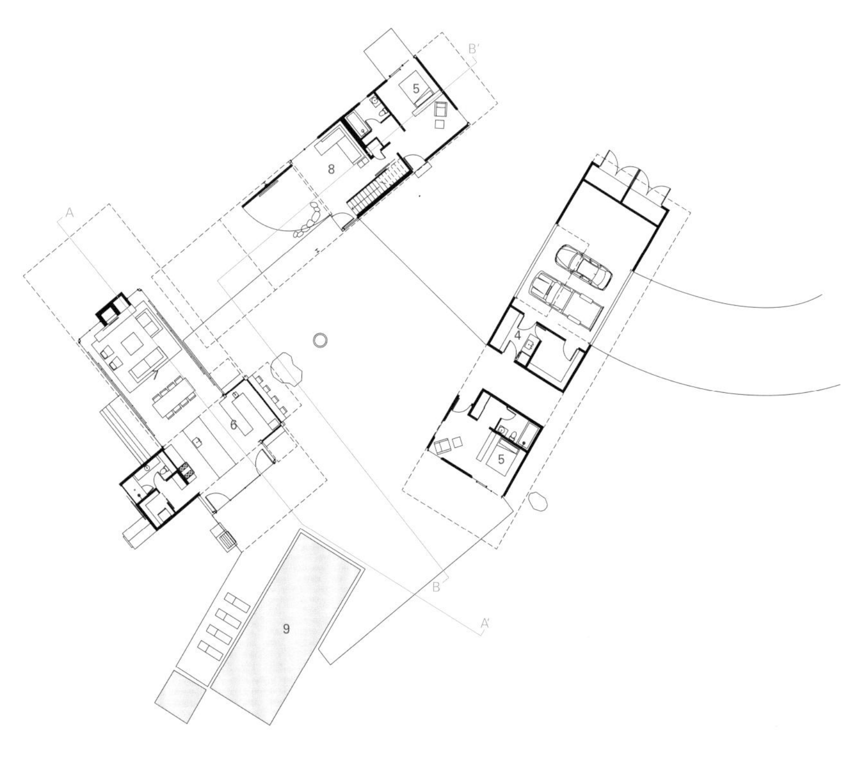

一层 ground floor

Studhorse

The sixty-mile-long glacial Methow Valley, in the northern Cascades of Washington state, is a special landscape, with a climate that ranges from hot, fire-prone summers to winters with heavy snow pack. It's a true four-season landscape – the landscape I grew up in. Our clients wanted to build a second home here that would be a kind of adventure home for them and their kids. They are great parents and always undertaking adventures as a mindful, deliberate way of developing memories as a family.

In this location, a house that's all about adventure is one that forces you to be outside and engage actively with the seasons. You have to go outside to get inside. So the house has inconveniences, but the clients and I see them as terrific moments, unforgettable moments. Adventure is about inconvenience in that it reaffirms and reminds you of where you live. I used to climb mountains, and while it may seem romantic, it's also uncomfortable. And scary. You're cold, hot, sore. Why would anyone do it, if they thought about it logically? But it's about engaging life vigorously.

The house is composed of four buildings, centered on a

central courtyard and pool. They are unattached but grouped together as a riff on the idea of "circling the wagons". It's like a little campground, and you go tent to tent. The materials are tough on the outside, because of the high-desert climate, but the inside is cozy, like getting into a sleeping bag – protected, warm, and dry.

Each building has an amazing, carefully composed view of the surrounding Studhorse Ridge and Pearrygin Lake. The "public" areas, such as the family room, kitchen, and bar, are grouped together in a structure that opens up almost entirely, with sliding glass windows all along the walls. The master bedroom, kids' bedroom, and den are in another building. The third is for guest rooms, to allow guests to have their privacy. A sauna is in the fourth building with a private, framed view looking out over the valley.

The site was completely empty when we began, except for that rock! It is a glacial erratic – a rock that glaciers drop as they recede – and it became the center point for the project. It's relatively close to the bar, so it becomes a place to sit, or put your drink. I thought of it as a big piece of furniture.

The design embraces family life and entertaining. For example, one wall with a TV can either face inside or open up to face the courtyard. If there's a game on, everyone can watch from the pool area. There's also a Coney Island aspect to the bar, where you push a button and the walls open up. It has turned out to be a particularly successful part of the architecture. Everybody hangs out naturally at the bar, and the fact that you can open it up is important to the inside-outside relationship we were seeking to achieve.

The wood siding used throughout the project was salvaged from an old barn in Spokane. The varying tones of the wood reveal its history and use. The glass is very important. I have never liked being in a room that doesn't feel like part of the outside. I always like to take the amount of glass I'm allowed per code and, rather than sprinkling it around like pepper, concentrate it in one place to make a vivid connection to the landscape. Olson Kundig Architects

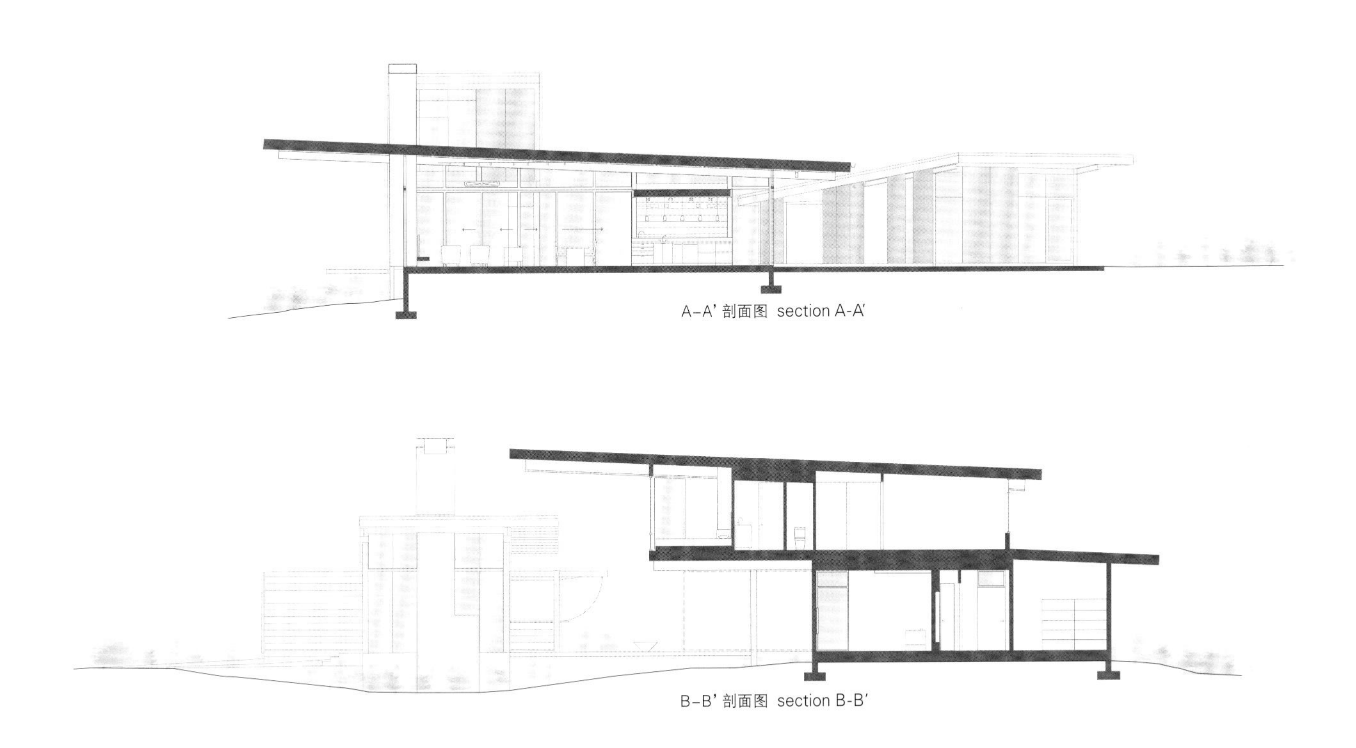

A–A' 剖面图 section A-A'

B–B' 剖面图 section B-B'

瓦赫特贝克周末度假屋

GAFPA

瓦赫特贝克的这个周末度假屋位于一排联排别墅的最末端，这里是弗兰德斯郊区典型的带状风格房屋到开阔农田的过渡区域。这个地块非常独特，黄色的连翘属植物丛生，从山坡向下蔓延到Moervaart河，为设计过程提供了主旨。

房子略微高于地面，利用细长的预制混凝土T形构件进行架空，这样可以尽可能多地俯瞰这里的美景。由于采用了这种轻型地基，这里的土壤基本上没有与建筑接触，这意味着可以在附近种植树木作为天然防晒的手段，甚至在房屋施工前就可以着手进行。

这间房屋全部朝向田野，堪称释放日常都市生活压力的理想所在。这个U形建筑靠街道的外墙基本封闭，只有个别狭长的洞口安装了多孔金属网百叶窗。相比之下，内墙完全开放，整个墙面都安装了玻璃，可以俯瞰农田，与周围的环境融为一体。

两个起居空间定义了平面布局，它们各自占据U形结构的一端，各自都有带顶露台。儿童卧室、浴室和存储间沿外墙布局设计，利用几个滑动门和橱柜与开放式走廊隔开。从二楼的开放型主卧沿着螺旋楼梯走下来便是一楼的厨房。

整个建筑被设计成一个木结构。一个隔热的预制木框架搭建在混凝土板上。承重内墙和地板结构是用层压交叉木材搭建的；外墙为木框架结构，用纤维板隔热，外用胶合板装饰。

立面的设计错落有致，透明和不透明构件穿插使用。幕墙由胶合板与三层玻璃构成，安装在轻型U形铝框架中，沿露台自然展开，把立面与外部区域整合在一起。

整个项目的结构和建筑构件仍然清晰可见。建筑接合处并不追求复杂的覆层或昂贵的材料，而是以一种合理的方式呈现出来。该建筑细部设计很简单，在建筑各层结构中的接合处都是显而易见的，所以无论是施工材料还是施工上的小瑕疵都是要尽量避免的。该项目的空间价值就在于它设计合理、层次清晰以及非常宜居。

Weekend House Wachtebeke

The last element in a row of detached houses, the Weekend-house in Wachtebeke occupies the transition from suburban

Flanders' characteristic ribbon development into the wide open farmlands beyond. This unique location, on a plot overgrown with yellow forsythia sloping down to the river Moervaart, provided the leitmotiv for the design process. The house is lifted up slightly, held aloft by slender prefabricated concrete T-elements, occupying the landscape as modestly as possible. This lightweight foundation leaves the soil largely untouched, which allowed for a tree to be planted close to the building as a means of natural sun-screening – the first work to be done, before construction had even started.

As a place for retreat from everyday city life, the house is entirely oriented towards the fields behind it. The U-shaped building presents its largely closed outer facade to the street, scarcely broken up by expanded metal shutters covering narrow openings. The inner facades, by contrast, are opened up completely - glazed along their full length, overlooking the farmlands, bringing the landscape inside.

The floor plan is defined by two living rooms, one in each of the U's arms, each with their own covered terrace. The children bedrooms, bathroom and storage spaces are organised along the outer wall, separated from the open corridor by a rhythm of sliding doors and cupboards. A spiral staircase descends from the open master bedroom at the first floor, and defines the kitchen area.

The whole building is conceived as a timber structure. A prefabricated and insulated wooden frame rests upon the concrete slabs extending from the ground. The load-bearing interior walls and floor structures are executed in cross-laminated timber; the outer walls are constructed as wooden frames, insulated with cellulose and finished with plywood panels.

The facade is designed on a strict rhythm, extending across transparent and opaque elements. The curtain wall of plywood panels and triple glazing is fit into a light frame of aluminium U-elements, which is continued along the terraces – extending the articulation of the facade to incorporate the outdoor spaces.

Throughout the project, structure and building components are left visible. Architectural articulation is not sought in intricate cladding or expensive materials, but in a logical means of construction. Simple detailing, in which the different layers of the building are openly articulated, results in an architecture which tolerates minor flaws in its construction materials or execution. The spatial value of the project lies in its logic, its clarity, and its inhabitability.

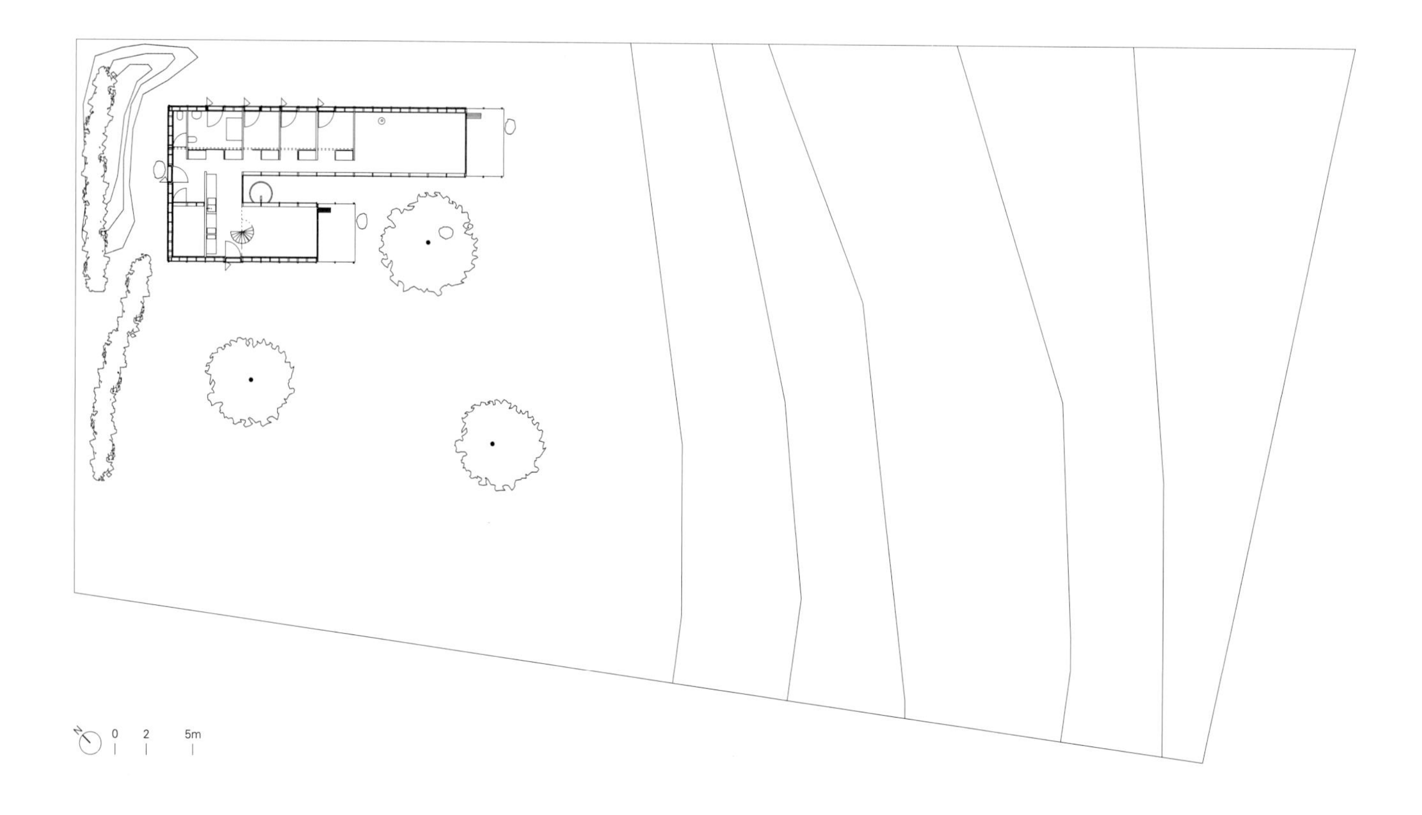

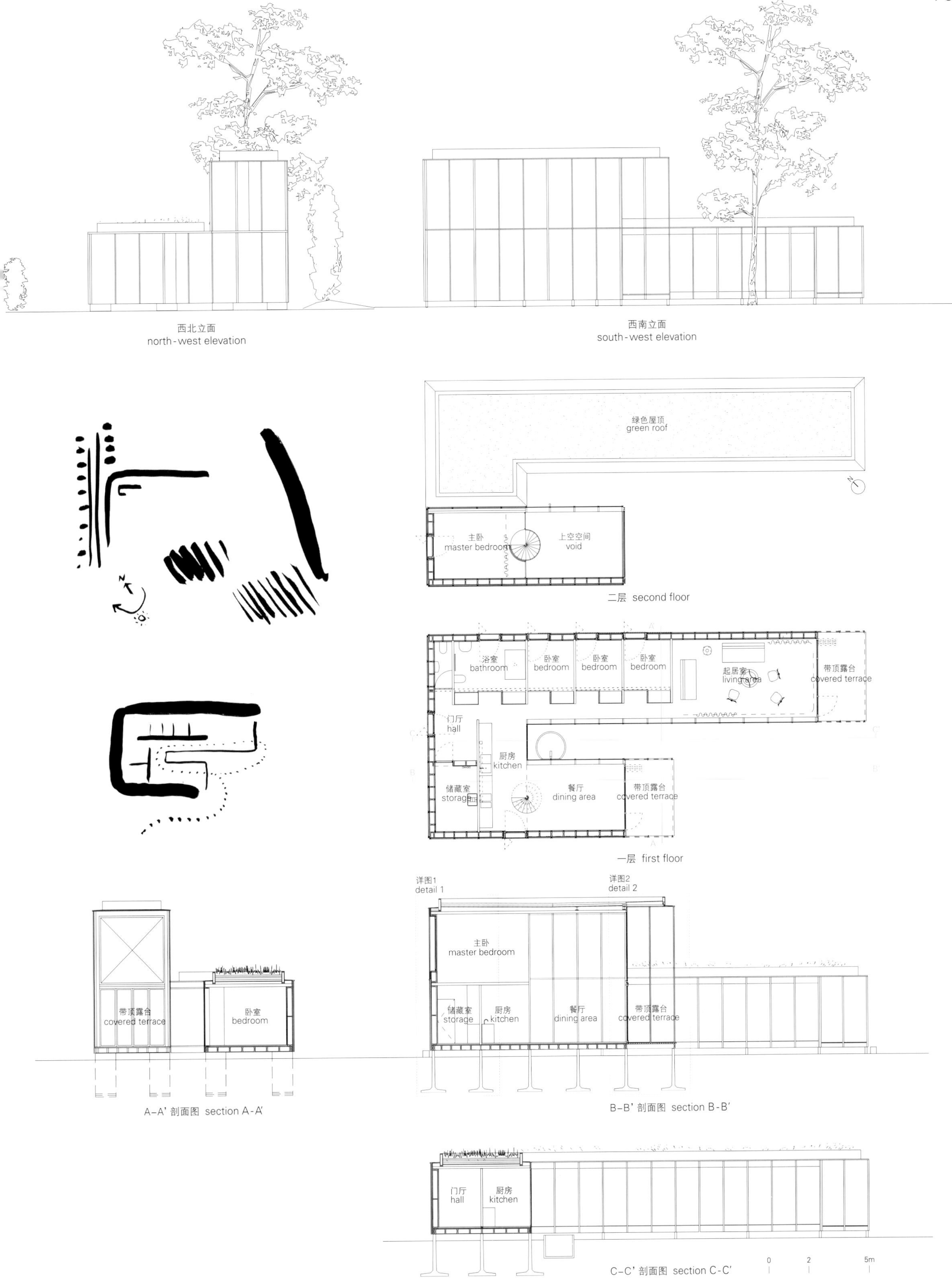
西北立面
north-west elevation
西南立面
south-west elevation
绿色屋顶
green roof
主卧
master bedroom
上空空间
void
二层 second floor
浴室
bathroom
卧室
bedroom
卧室
bedroom
卧室
bedroom
起居室
living area
带顶露台
covered terrace
门厅
hall
厨房
kitchen
储藏室
storage
餐厅
dining area
带顶露台
covered terrace
一层 first floor
详图1
detail 1
详图2
detail 2
带顶露台
covered terrace
卧室
bedroom
A-A' 剖面图 section A-A'
主卧
master bedroom
储藏室
storage
厨房
kitchen
餐厅
dining area
带顶露台
covered terrace
B-B' 剖面图 section B-B'
门厅
hall
厨房
kitchen
C-C' 剖面图 section C-C'
0
2
5m

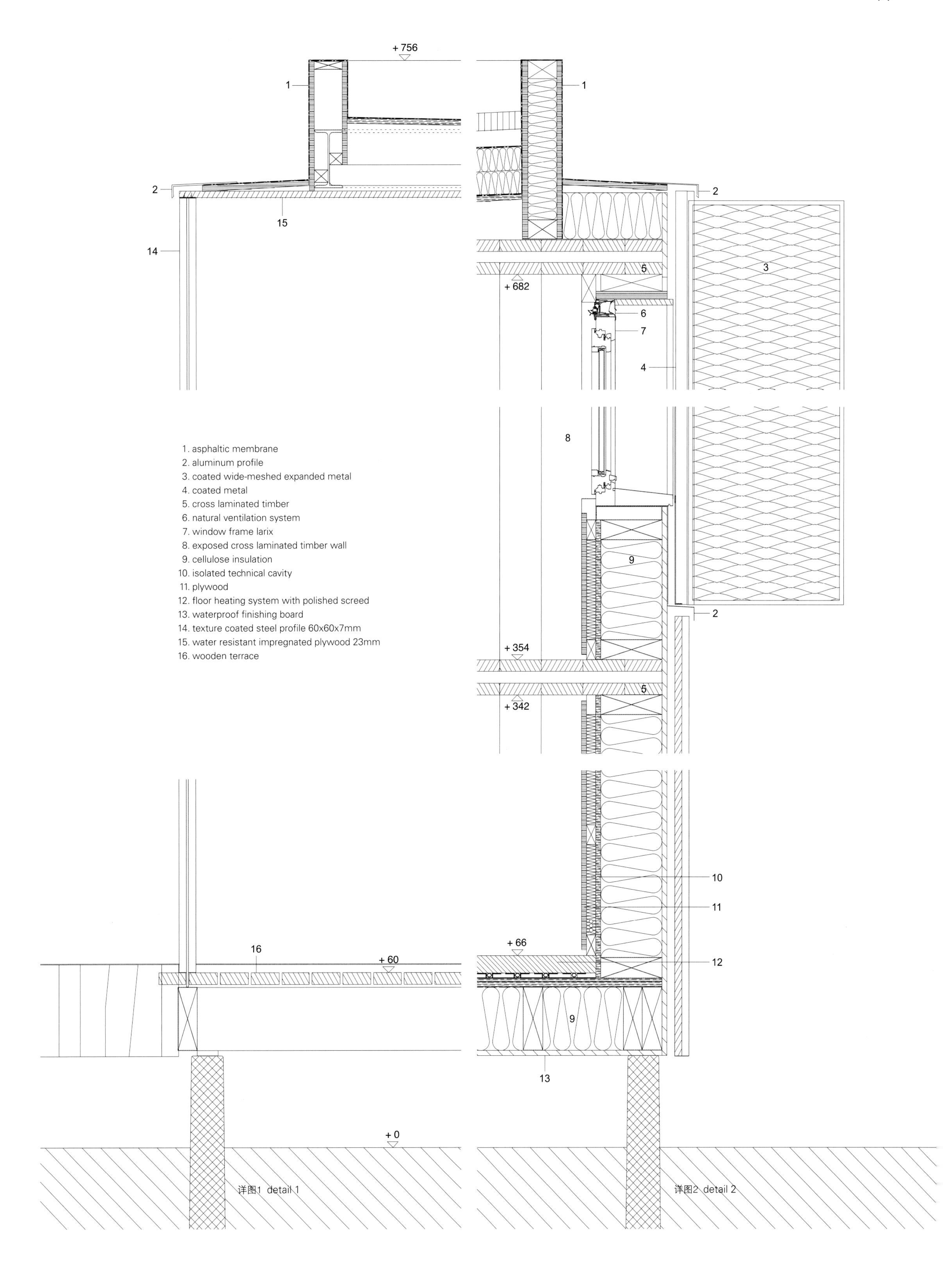

详图1 detail 1

详图2 detail 2

项目名称：Weekend House, Wachtebeke / 地点：Wachtebeke, Belgium / 建筑师：GAFPA / 总承包商：VH-building / 木工：Lab 15 / 外部细木工：Wood Frames
金属部件承包商：Tinel Interiors / 外部施工：Patrick 't Hooft / 工程师：Arthur De Roover / 用地面积：3,100m² / 建筑面积：160m² / 设计时间：2011 / 竣工时间：2014
摄影师：courtesy of the architect

大城府撒拉酒店

Onion

大城府撒拉酒店是一家拥有26个客房的精品酒店，位于湄南河沿岸风景如画的泰国旧都。沙旺寺由大城王国的第一代帝王建于公元前1353年。从撒拉酒店的餐厅和湖景套房都能看到它。撒拉酒店的主要入口位于乌通路上一棵撒拉树的旁边。长长的砖头外墙上有一扇铁门，将我们带入有着低矮天花板的接待处和双倍体量的画廊。这里最主要的特征就是Onion设计的古老木门，它位于透明的镜子中间，向外面的庭院敞开着，若干平行的多曲面砖墙使庭院变得越来越窄。这些墙围合出了天空的图景。这个主要循环通道的独特之处就在于不断变化的光影。两端弯曲的阴影通常在早上11点钟在地面上会合。它们在一天的不同时间内不断地改变空间的氛围。

在砖墙庭院里无法看到湄南河和沙旺寺的全景，我们只能穿过撒拉酒店餐厅走向河畔平台上才能欣赏到。在这里，我们能看到撒拉酒店的另一侧立面，山墙房屋的白墙、阶梯式平台的布局以及水边的露台。撒拉酒店的布局是12个私人住宅的融合，这些住宅相应地分布在河边面积有限的L形土地上，室外空间才是该项目的焦点。室外空间展现了当地的砖墙工艺，它将白色砖墙的简单与整洁融为一体，还解决了一年一度的防洪问题，这些设计都受到了印度拉贾斯坦邦的月亮深井的启发。

大城府撒拉酒店中的阶梯式平台可以被水淹没，人们可以沿着阶梯从一层高的平面往下走到与水面齐平的地方。四棵开红色花的树叫作“jik”，栽种在主平台旁边，以标示出户外酒吧的位置。沿着狭窄的河岸，Onion栽种了一排叫作“krading-nangfa”（泰语中的意思是“天使的铃铛”）的热带植物。这种植物开出铃铛形状的花，开满花朵的树枝对着河面向下垂着，形成了一条长长的花树隧道，芬芳而优美。随着时间的流逝，大城府撒拉酒店将日臻完善。台阶的图案也可以被解释为大城府建筑设计元素的一部分。建筑师借鉴了普泰萨万佛塔的尖端逐渐变小的形状，按照相同比例缩小再重新设计，用在了撒拉酒店的各种物品上，其中包括室内与室外的墙壁、水平与垂直的墙面以及家具和枕套的设计。这种重复的设计理念体现了建筑师从功能和装饰角度对当代泰国内涵的思考。

Onion更加关注施工细节以及室内的设计。铃铛形状的花岗岩灯是撒拉酒店餐厅的主要设计风格，这些灯在当地一家工厂专门定做。挂钩的结构模式基于网格系统设计而成，但是设计看上去充满动感，这是因为将电缆图案从直线变成Z形的重物位置都非常精确。一幅在泰国人眼中代表力量的老虎向前腾空一跃的图画被刻在木床的床头上。一种被称为“luk—mahuad”的泰式装饰物的剖面经过重新设计成为浴室柜和床的一部分。大城府撒拉酒店体现了设计师对形式与定制化建造的关注。

来到撒拉酒店的顾客的个人隐私可以通过酒店的环形设计得到保护。建筑师决定采用可以让客人直接通往卧室的走廊，而不是一连串的楼梯。每间卧室都不一样，而且至少有三个房间可以直接通向由白色大理石做成的游泳池。最舒适的房间是一个较小的卧室，带有独立私人露台和可隐藏的供孩子使用的坐卧两用沙发。从一个较宽敞的房间看不到河岸的景色，但是它的纵向墙壁朝着游泳池的纵向方位。再上一层，在画廊上方的房间可鸟瞰砖墙庭院。在连桥房间也可以鸟瞰游泳池和花园。每个房间的这种独特的设计使得客人们希望能再次光临撒拉酒店。

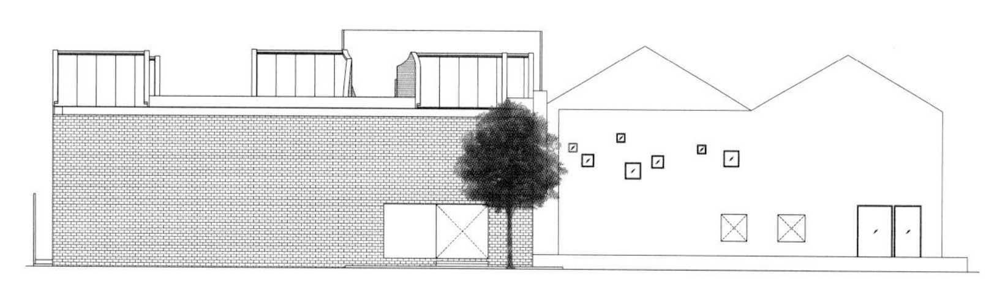

北立面 north elevation

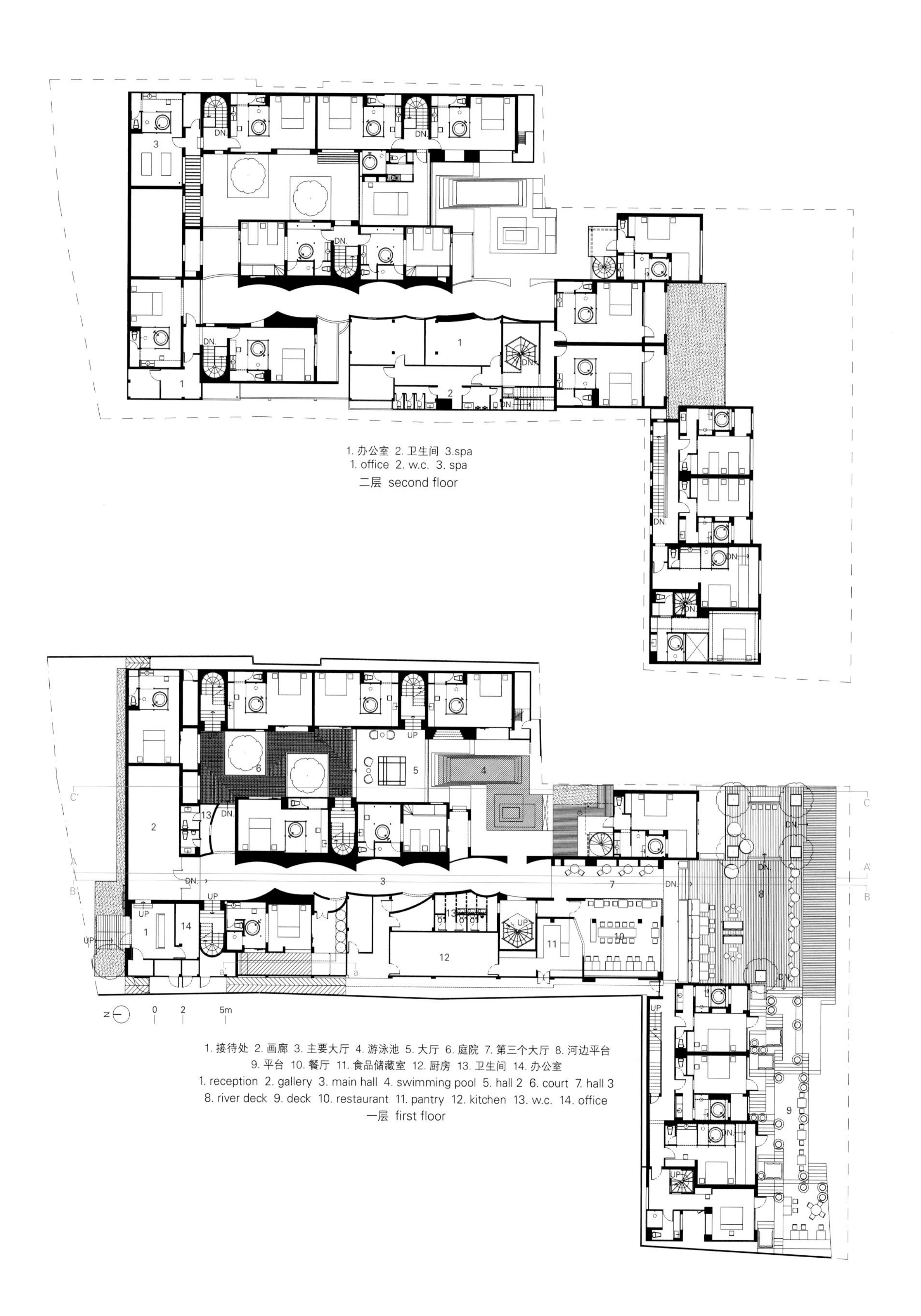

1. 办公室 2. 卫生间 3.spa
1. office 2. w.c. 3. spa
二层 second floor

1. 接待处 2. 画廊 3. 主要大厅 4. 游泳池 5. 大厅 6. 庭院 7. 第三个大厅 8. 河边平台
9. 平台 10. 餐厅 11. 食品储藏室 12. 厨房 13. 卫生间 14. 办公室
1. reception 2. gallery 3. main hall 4. swimming pool 5. hall 2 6. court 7. hall 3
8. river deck 9. deck 10. restaurant 11. pantry 12. kitchen 13. w.c. 14. office
一层 first floor

Sala Ayutthaya Hotel

Sala Ayutthaya is the twenty-six-room boutique hotel, right across a most picturesque site of the old capital of Thailand along the Chao Phraya River. Phutthai Sawan Temple was built in 1353 AD by the first monarch of Ayutthaya Kingdom. It becomes the view of the restaurant and the riverfront suites at Sala. The main entry of Sala is next to a Sala Tree on U-thong Road. It is a single iron door on a long brick facade, leading us to the low wooden ceiling reception and the double volume art gallery. Within this space, the dominant feature is an antique wooden door that Onion designs the framing for. It is placed between the transparent mirrors, opening to the exterior courtyard, narrowed by the paralleled brick walls of multi-curved geometries. They frame the image of the sky. What is unique about this main circulation is the constantly changing shadows. The curved shadows from the two sides normally meet on the floor at about eleven o' clock in the morning. They transform the atmosphere of the space at different times of the day.

The panoramic view of Chao Phraya River and Putthai Sawan Temple cannot be seen from the brick walls courtyard. It can only be experienced when we walk pass Sala Restaurant towards the riverfront deck. At this location, we see another facade of Sala, white walls of gable houses, the arrangement of step decks and terraces along the waterfront. Sala layout is a compound of twelve private residences, proportionately packed within the limited area of L-shape land, leaving the focal points of the project to be the outdoor spaces. They exhibit the local craftsmanship of brickworks, juxtaposed with the simplicity and neatness of the white walls and a solution to the problem of annual flooding, inspired by Chand Baori

Step Well in Rajasthan, India.
At Sala Ayutthaya, the step decks are designed to be flooded. They lead us down from the one-storey high platform to the same level as the river. Four red-flower trees called jik are planted next to the main deck in order to signify the location of the outdoor bar. Along the narrowed river bank, Onion plants a row of tropical trees named krading-nangfa (its literal translation is the "angel's bell" tree). Their bell-shaped flowers will eventually blossom. Their branches will suspend themselves down towards the river and form a long tunnel of fragrant trees. Sala Ayutthaya will be more complete with age.
The pattern of steps can also be explained as an architectural element of Ayutthaya architecture. The architects have borrowed a reduced size of a corner of the Phutthai Sawan Stupa, scaled and redesigned it to frame various objects at Sala. This includes the interior and the exterior walls, the vertical and the horizontal planes, the furniture and the pillow cases. Such a repetition marks the architects' concern about what contemporary Thai is meant, in both functional and decorative senses.
Onion places much attentions on the construction details and interior designs. The bell-shaped granite lamps are the main features at Sala Restaurant. They are custom-made by a local factory. The structural pattern of the hangers is based on the grid system, but the design looks dynamic because of the precise positions of the weights that change the pattern of electric cables, from a straight to a zig-zag pattern. The image of a tiger leaping forwards, meaning strength in Thai's belief,

is engraved upon the head of the wooden bed. The section of a Thai ornament called luk-mahuad is redesigned as parts of bathroom counters and beds. Sala Ayutthaya projects a character of the architects who are interested in form and customisation.

The privacy of Sala guests is secured through the circulation designs. The architects decide to use a single load corridor much less than a series of staircase that directly lead the clients to their own bedrooms. Each bedroom is always different from the other. There are at least three rooms that have the direct access to the step swimming pool made of white marbles. The most cozy one is a smaller bedroom that has its own private terrace and a hidden daybed for children. A more spacious room does not have the river view, but its longitudinal wall is facing the longitudinal side of the swimming pool. On the upper floor, the room above the gallery has the bird-eye-view of the brick walls courtyard. The bridge room has the bird-eye-view of the swimming pool and the garden courtyard. These special characters of each room make it exciting for the guests to revisit Sala.

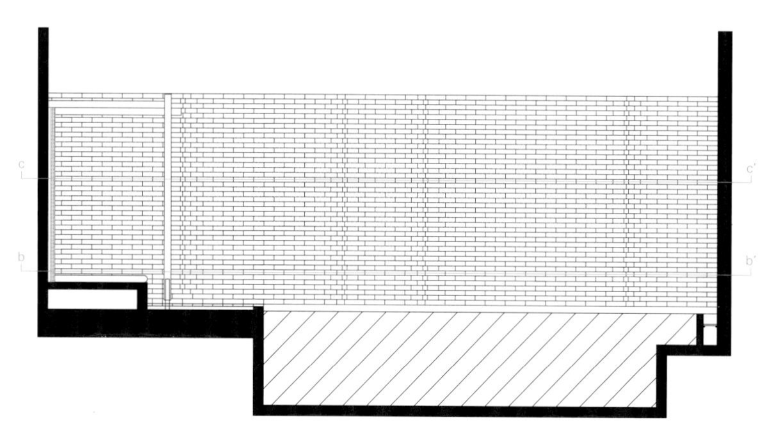

a-a' 详图 detail a-a'

项目名称：Sala Ayutthaya Hotel
地点：Phra Nakhon Si Ayutthaya, Thailand
建筑师：Siriyot Chaiamnuay, Arisara Chaktranon _ Onion
室内设计师：Siriyot Chaiamnuay, Arisara Chaktranon _ Onion
文本：M.L. Chittawadi Chitrabongs
面积：3,500 sq.m
竣工时间：2014.8
摄影师：©Wison Tungthunya (courtesy of the architect)

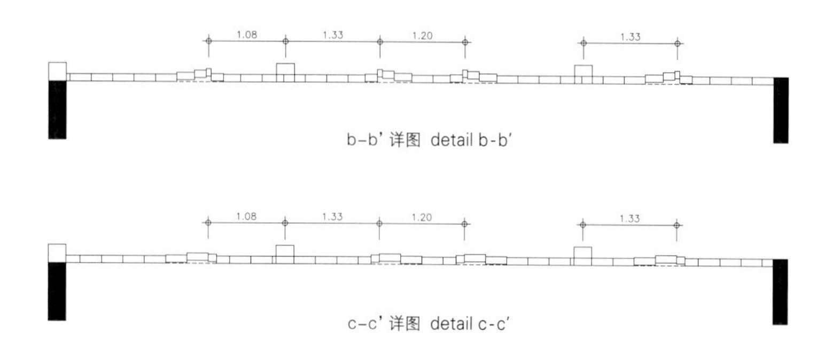

b-b' 详图 detail b-b'

c-c' 详图 detail c-c'

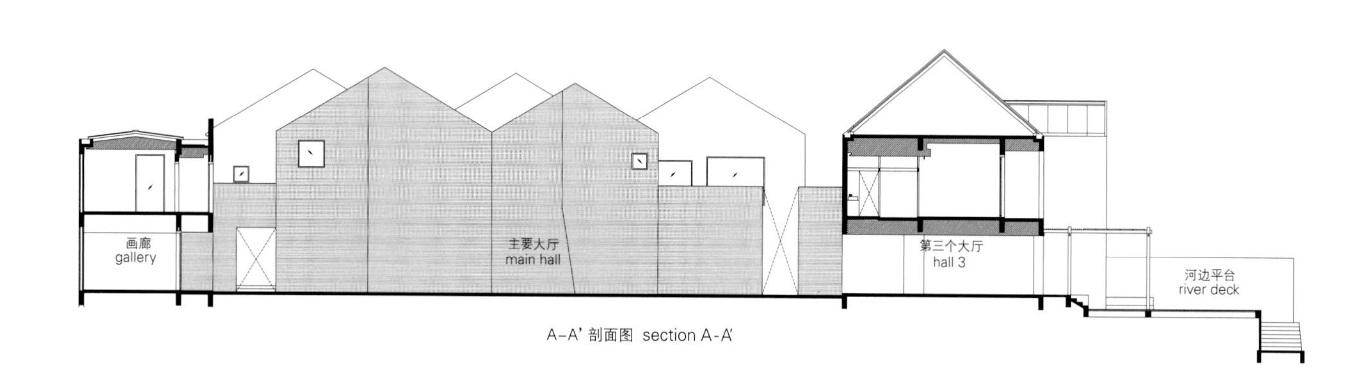

A-A' 剖面图 section A-A'

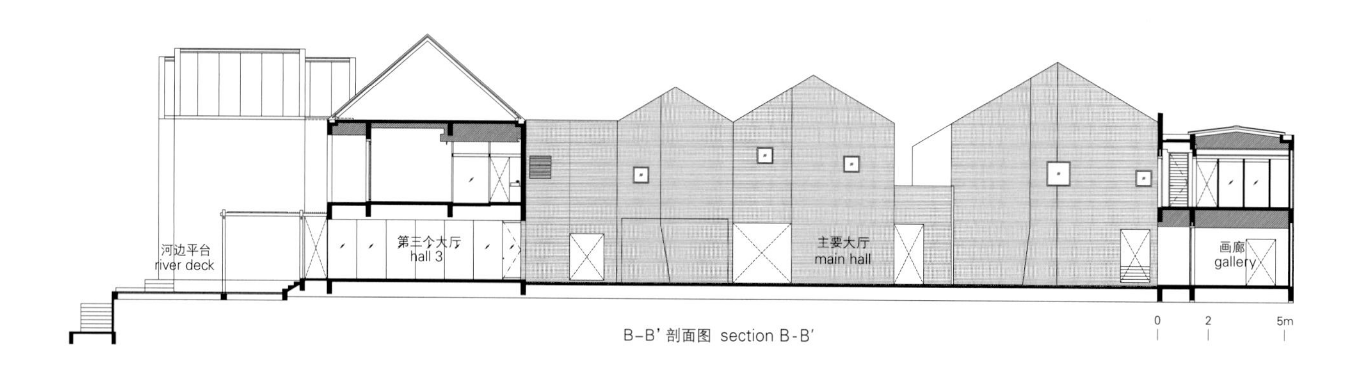

B-B' 剖面图 section B-B'

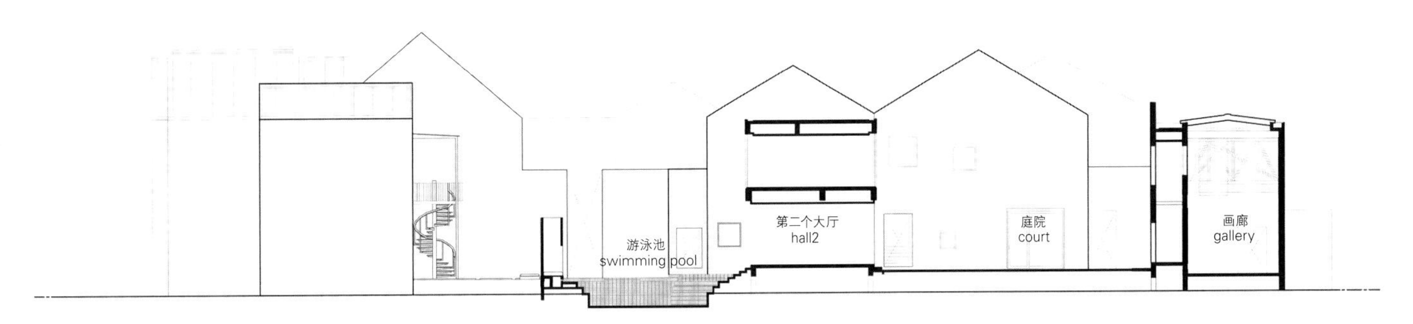

C-C' 剖面图 section C-C'

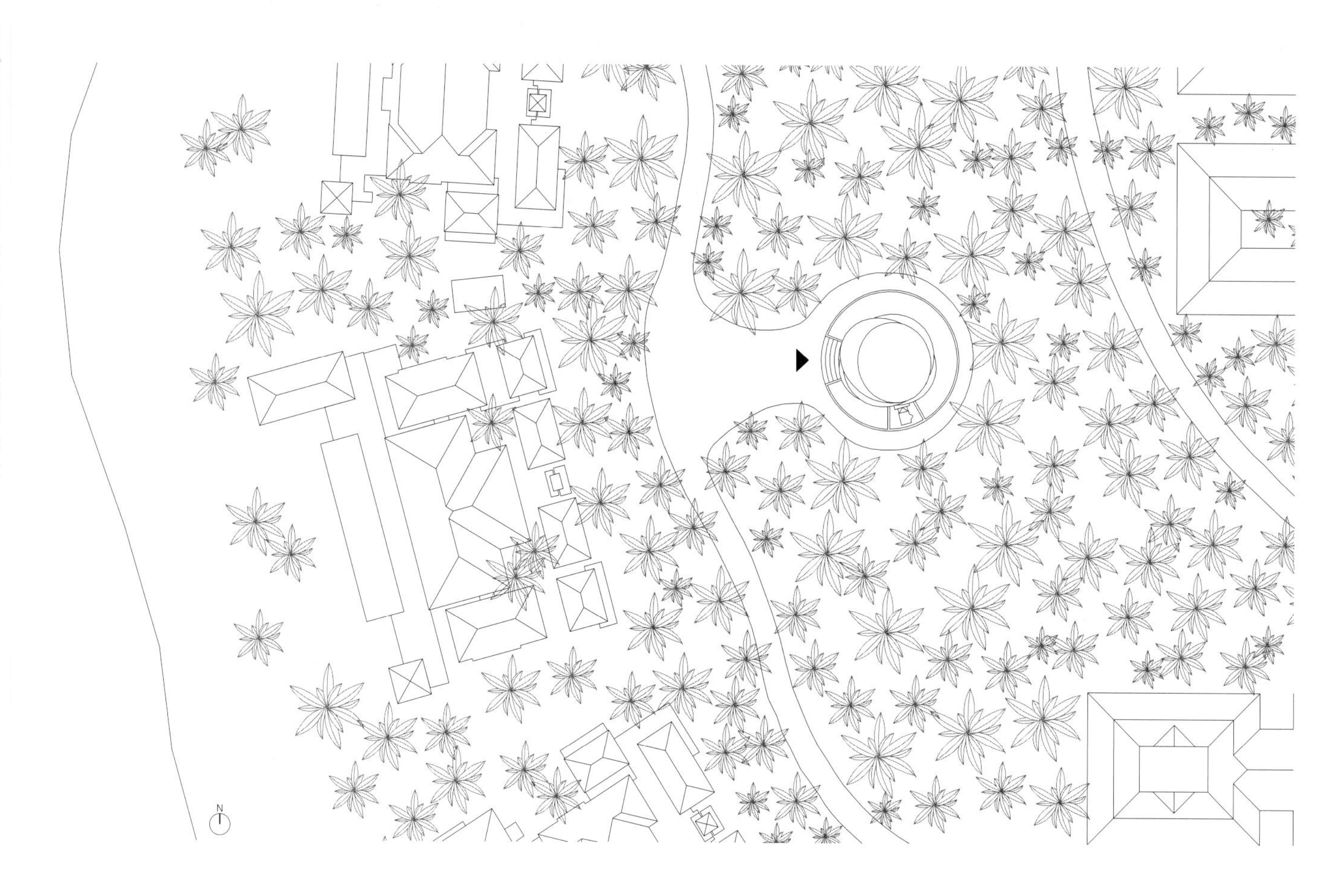

Tavaru餐厅酒吧

ADR

Tavaru餐厅大楼是马尔代夫地区独一无二的地标性建筑，它给予到访者的体验也是马尔代夫独一无二的。其建筑设计有别于岛上的其他建筑——它现代、超前，又颇"浑然天成"。此外，它将打造一个拥有合适微气候的环境，特别适合陈列及窖藏世界上最优质的红酒，同时搭配口味绝佳的日本料理供游客品尝。Tavaru不重数量，只重品质，以最高品质来诠释"私享一刻"的生活哲学。

设计这座建筑的最大挑战在于不仅要使它与岛上的其他建筑相和谐，同时还要保持其自身的独特外观及张力。

期间，建筑师遇到的技术挑战是如何处理结构的基础部分以及如何在岛上的水文地质条件下建造一座高层建筑；还有一个更大的挑战是如何对这样的建筑进行施工，这是马尔代夫建筑中的创新型大楼，结合了传统的当地建筑方法和典型的现代欧洲施工技术与材料。

与岛上的其他传统"Tavaru"风格的建筑物相比，这座高23m、造型独特的建筑足以吸引全岛的目光。该建筑摒弃了当地传统的天然材料，在表面采用了钢筋混凝土单体结构，垂直的桶状构造作为支撑结构，外围安装了由预制楼板制成的楼梯，盘旋而上。建筑主体外再罩以浅灰色的半透明幕墙。酒窖占据较低的两层。楼上是餐厅及其附属空间。最上层是马尔代夫唯一可以360度尽览岛上美景的酒吧休息厅。

Tavaru Restaurant & Bar

This tower restaurant is unique in the context of the Maldives region; it offers visitors experiences that are not to be had anywhere else in the Maldives. Its architecture will set it apart from all the other buildings on the island – it will be modern, futuristic, finely "raw". Last but not least it will create an environment and suitable micro-climate for storage and presentation of a unique collection of the world's best wine vintages, combined with extraordinary Japanese experiential cuisine. Not quantity - but the highest attainable quality in the spirit of the "Private Moments" philosophy.
The greatest challenge in terms of architecture was to achieve a harmonious symbiosis with other buildings on the island, while maintaining a visible contrast and tension.
A technical challenge was to resolve the structural founda-

tions and building a high-rise building in the hydro-geological conditions of the island; an even greater challenge was then to plan the actual construction of such a building, which is one of a kind in the Maldives – innovative, combining traditional and local building methods with techniques and materials which are typical of modern European construction. In contrast to the other buildings on the island, which are in the traditional "Tavaru" style, the present structure, due to its height of 23 meters and specific construction, dominates the island. Instead of traditional natural materials, reinforced concrete monolith was used on the face of the building; it serves as a bearing tube, around which a console staircase made of individual prefab steps is spiralling. All elements on the external perimeter are veiled in a light grey semi-transparent curtain. The wine cellar takes up two lower floors. The remaining floors serve the restaurant facilities and the restaurant itself. The last floor houses a lounge with a vista of the whole island – the only panoramic view to be had in all of the Maldives.

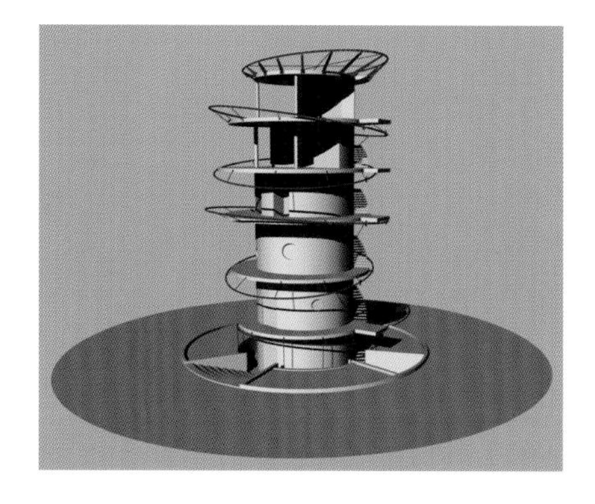

项目名称：Tower/Tavaru
地点：Velaa Private Island, P.O. BOX 2071,
Noonu Atoll, Republic of Maldives
建筑师：ADR
作者：ADR s.r.o._Mgr. A. Petr Kolar, Mgr. A. Ales Lapka
合作方：Ing. arch. Ivo Bartonek, Ing. arch. Anna Vildova
总承包商：Techo, a.s., Kontis Praha s.r.o.,
Aima Construction Company Pvt. Ltd., GAST-PRO s.r.o.
面积：25.082m²
设计时间：2010 / 施工时间：2011—2014
摄影师：©Archive ADR

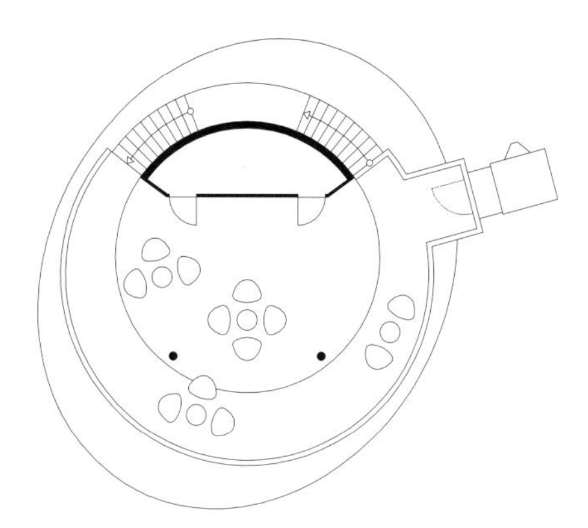

五层 fifth floor

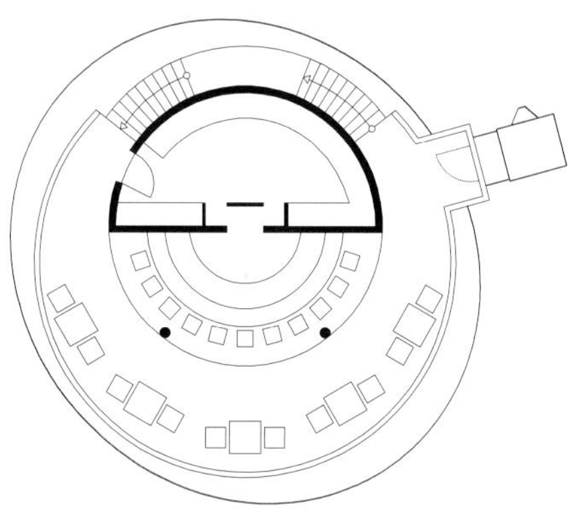

四层 fourth floor

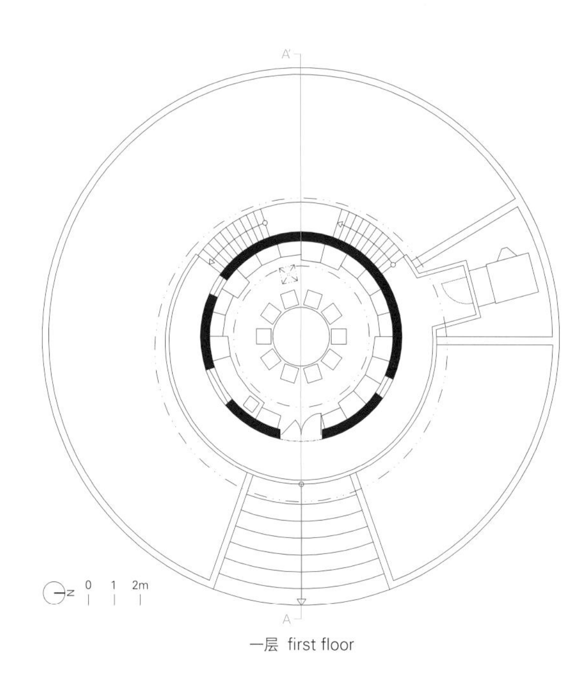

一层 first floor

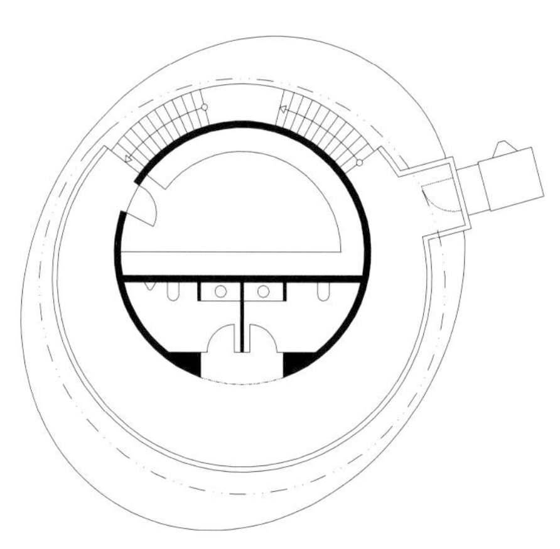

三层 third floor

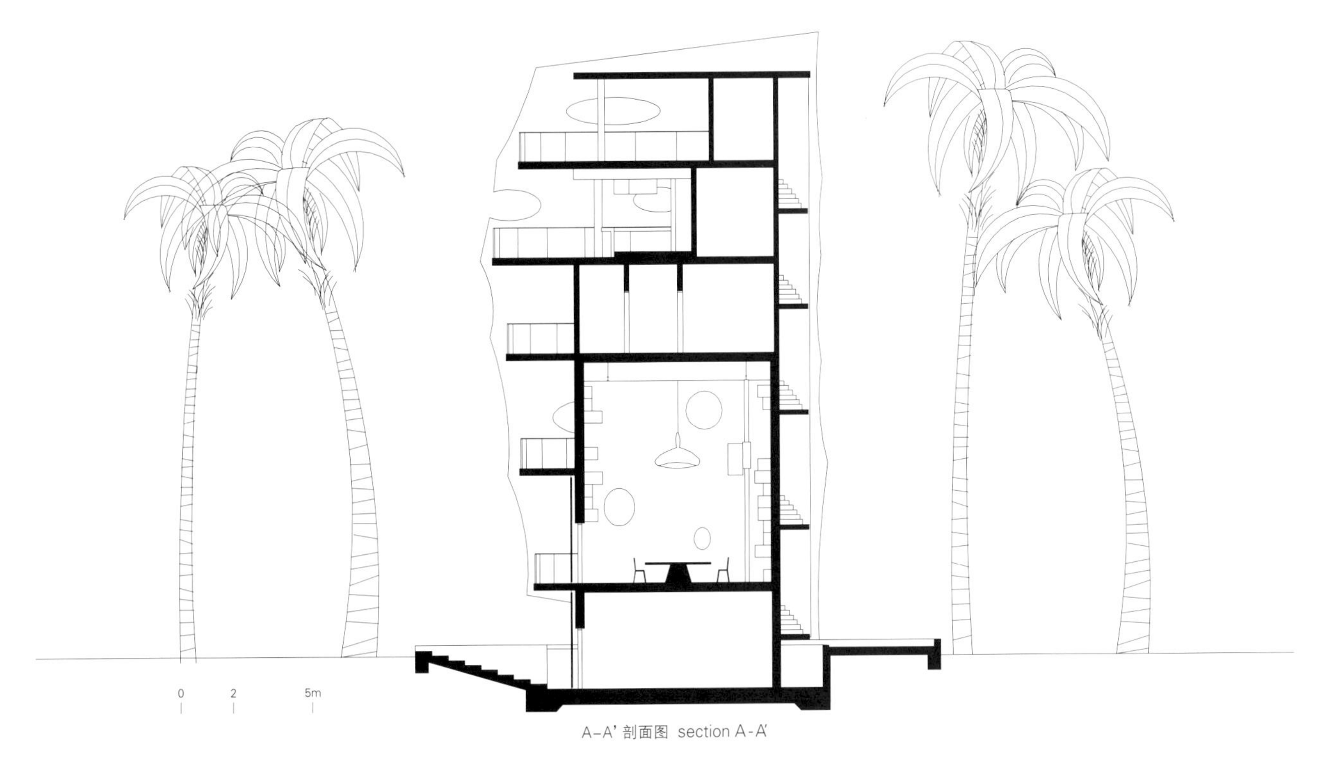

A–A' 剖面图 section A-A'

玛尔·阿登特罗酒店

Miguel Ángel Aragonés

第一次实地探访这个地块，进入这片沙漠，看到不远处清澈透明的海水沿着水平线流动，在炎炎烈日下，我感受到了水的巨大动力。这块土地位于海岸线的中央，包罗万象，考虑到这个世界创造了一片与大海毗邻而居的沙漠，形成浑然一体的美好画面，设计师将它变成了一个包含了属于自己的独一无二的海和空气的盒子，这是地平线上所能描绘出的最纯粹、最简约的景观。另外，这片梦幻般的美景与人们心中的美发生了碰撞，于是它经受了建筑学的洗礼。除此之外，我也想画一幅属于自己的作品。

用敏感的心境去发现一系列平面空间，进而推动"感觉"的产生，我认为这就是建筑最伟大的美德所在。若能与周围的环境相融合，即达到空间与身体合而为一的境界，这种能力就会变得越发强大。在这个意义上，我想将地平线转换为前景。水是整个项目的关键，它环绕着建筑群，所有建筑体量都面朝大海，背对城市，繁华喧嚣的都市被远远地抛在身后。玛尔·阿登特罗酒店面朝大海，实际上是另一种意义上的麦地那。每个体量都包含其独立的内部空间，如同一个个孤岛，悬浮于水面之上。每个房间都有专属的绝佳海景视角，无人能够抗拒这无敌海景的魅力。

每个房间都是在工厂里建造的。我们在工厂里建造了整个内部结构，随后装进箱子将其穿越海洋运至目的地，最后由当地工人现场组装完成。没用几天，第一间房间就完成了，工厂机器的现代化程度及终身致力于建造的工人的技能决定了建筑的质量。在这里，没有即兴发挥的空间，但毋庸置疑的是，该建筑的房间是由智慧、想象力及工人的奉献精神共同塑造完成的。与此同时，我从德国和意大利制造商身上学到了许多知识，这些知识并不是在多年的学校或书本学习之后仅凭直觉就能知道的。我们的项目可以完全依靠这个过程得以实施，采用一个可以随意切分或添加的多功能模块，因此可以自由变换——或自成一体，或从属于其他结构。例如，该建筑的主要模块是一个阁楼，为了形成两个独立的房间，该阁楼可以被切分成两部分，道理就是这么简单。总而言之，模块可以是一套两居室或三居室，甚至是四居室的公寓；再添加两个或四个模块，一座房子就建造完成了。这里最重要的就是结构的多功能性，它可以完全在工厂预制，接着以温和的方式在现场建造。

Mar Adentro Hotel

The first time I visited the site and took in the desert and the diaphanous, clear water running along a horizontal line in the background, I felt the enormous drive of water under a scorching sun. This piece of land, located in the middle of a coastline dotted with "All Inclusives" would have to be transformed into a box that contained its own sea – practically its own air – given the happy circumstance that the universe had created a desert joined to the sea along a horizontal line. It was the purest, most minimalist landscape a horizon could have drawn. On either side, this dreamlike scenery collided with what humans consider to be aesthetic and build and baptize as architecture. I wanted to draw my own version, apart from the rest.

I believe that the greatest virtue of architecture is the generation of sensations through space on a series of planes that are found within the realm of sensitivity. I believe this capacity

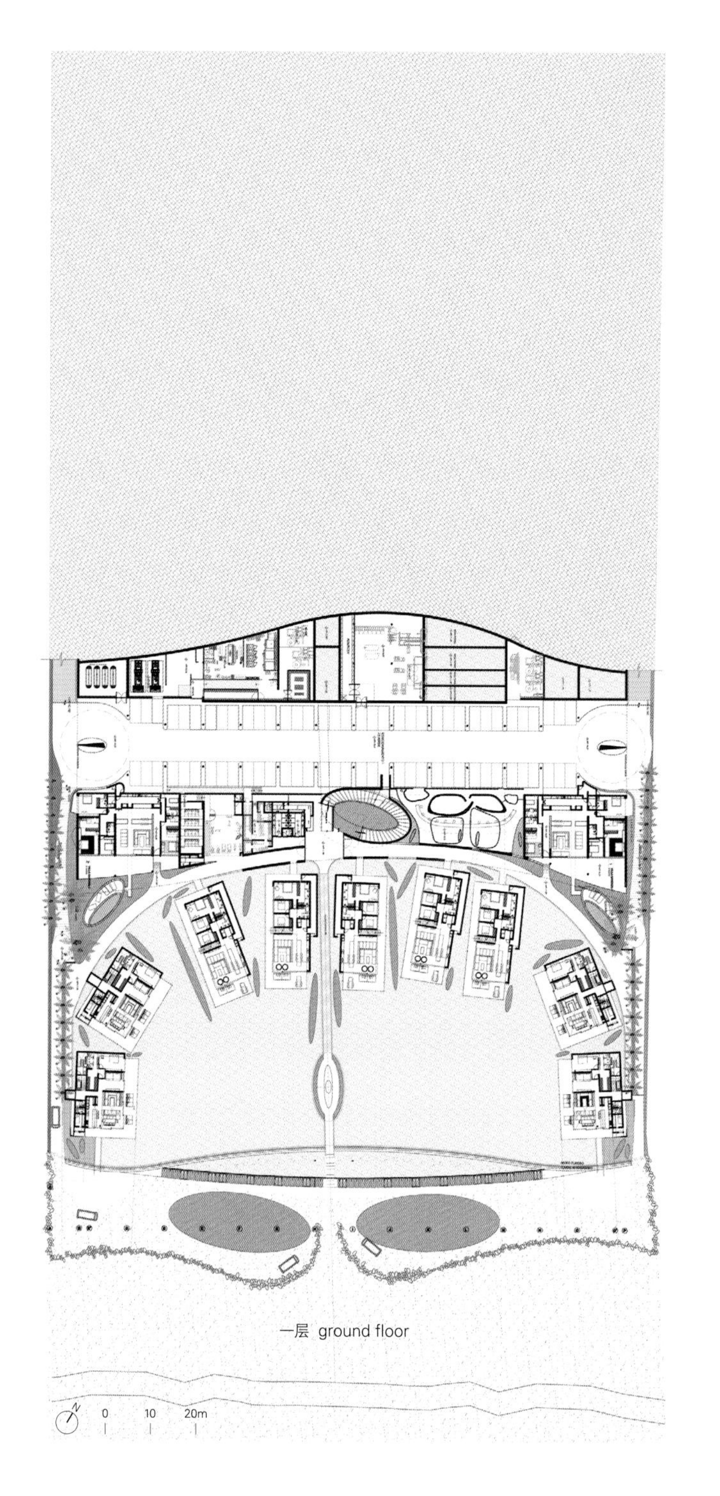

一层 ground floor

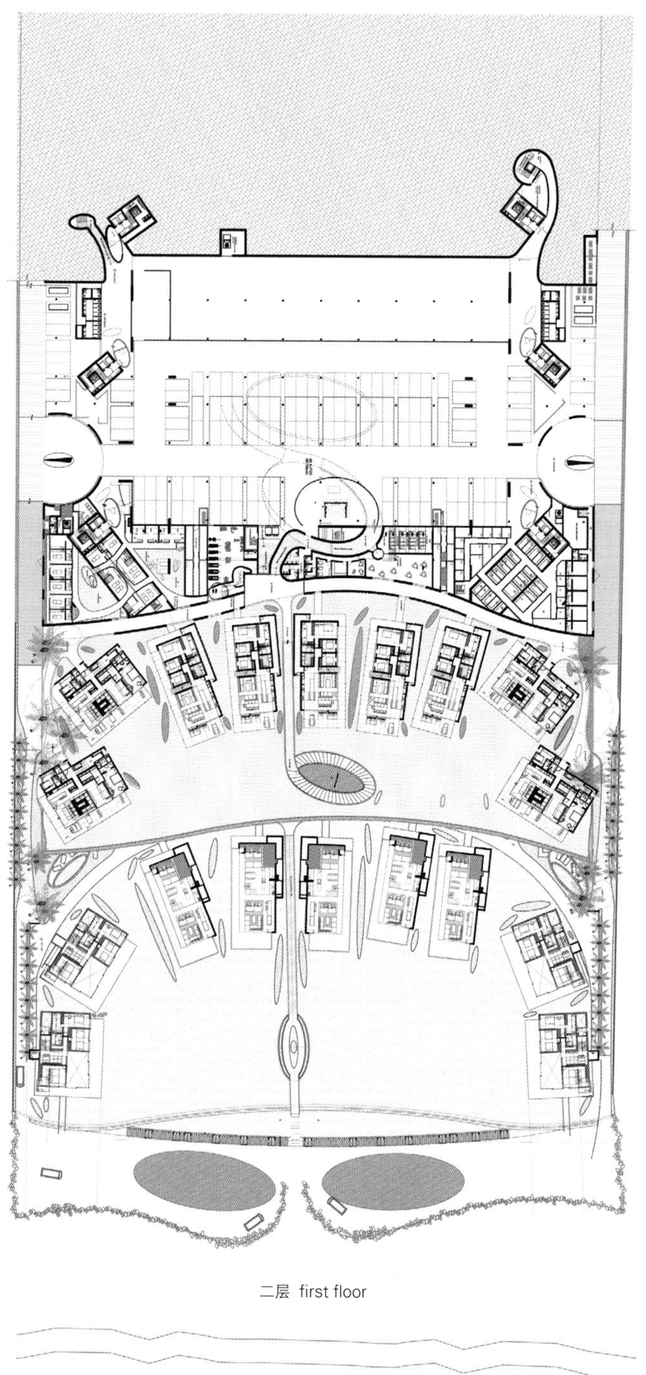

二层 first floor

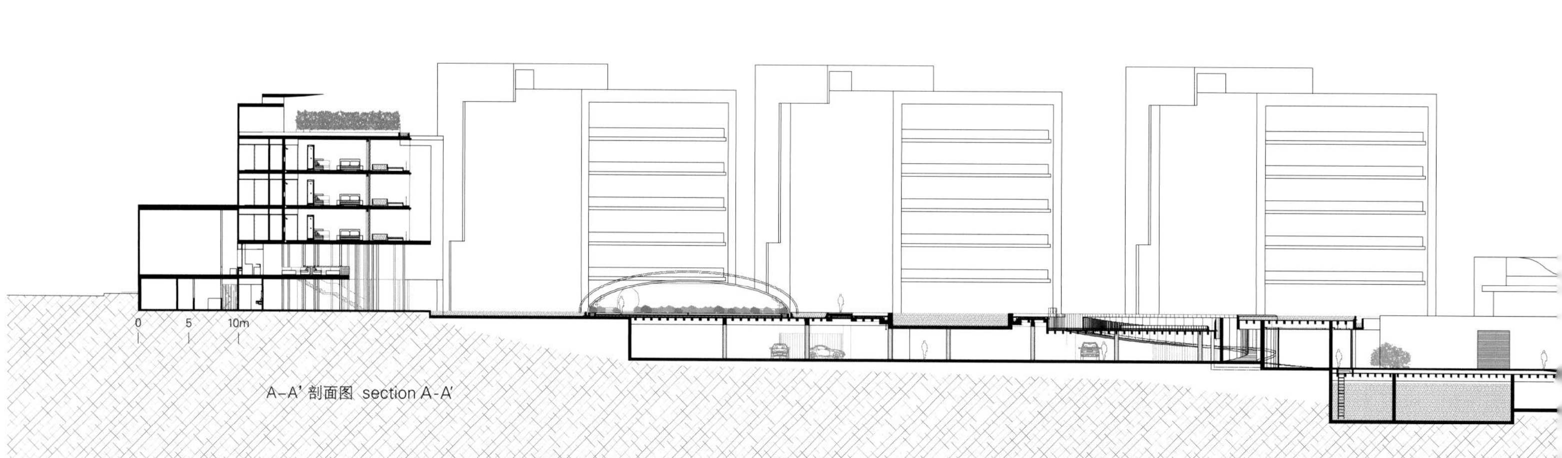

A–A' 剖面图 section A–A'

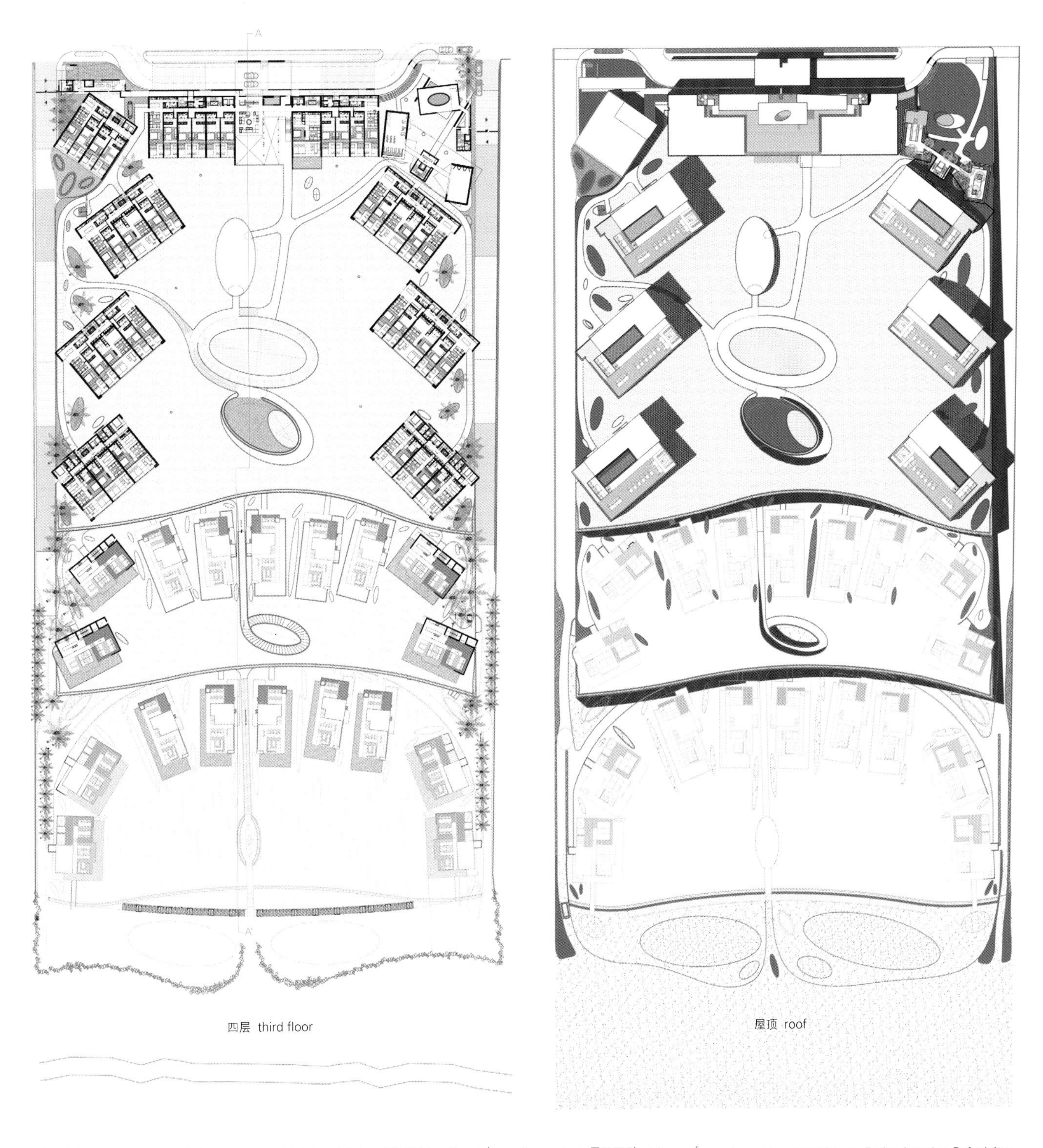

四层 third floor

屋顶 roof

项目名称：Mar Adentro / 地点：San José del Cabo, México / 建筑师：Miguel Ángel Aragonés / 项目团队：Miguel Ángel Aragonés, Juan Vidaña, Pedro Amador, Rafael Aragonés, Alba Ortega / 合作方：Ana Aragonés, Fernanda Kurth, Antonio Trinidad, Manuel de la O., Diego Amador / 结构工程师：José Nolasco / 特邀工程师：High Tech Services / 施工工程师：Jorge Flores, José Torres / 施工队工头：Severiano Torres, Roberto Torres / 照明工程师：Taller Aragonés, Ilumileds / 水力系统工程师：Swimquip / 智能系统工程师：Control 4. / 建筑面积：26,454.77m² / 总建筑面积：47,082m² / 材料：Concrete, Steel, Travertine / 设计时间：2012.1 / 施工时间：2014.11 / 竣工时间：2016.1 / 摄影师：©Joe Fletcher (courtesy of the architect)

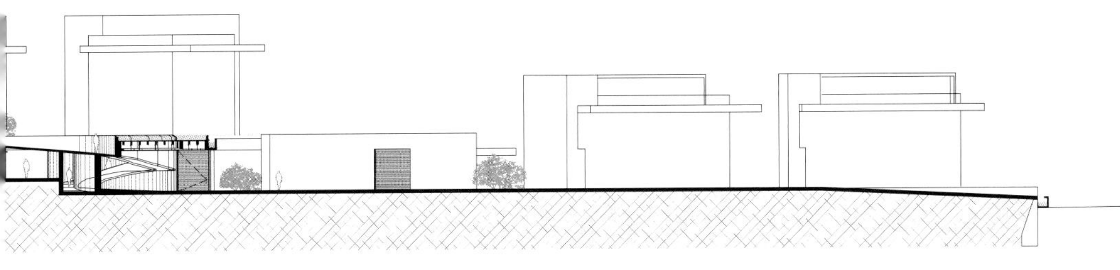

becomes still greater when your surroundings allow you to meld into them, forming thus part of your own space; in this sense, I wanted to take that horizon and bring it into the foreground. The water is an event that borders the entire project; all of the volumes open up toward the sea and turn their backs on the city, which is all that remains of the original surroundings, burdened by noise. Mar Adentro is a kind of Medina that opens out onto the sea. Each floating volume contains interiors that form, in turn, independent universes. Each room visually contains a piece of the sea; no one can resist gazing out at it.

Each room was built in a factory. We built the entire interior structure and sent it in boxes across the sea to its destination, where it was assembled on site by local hands. In a question of days the first room was ready, of a quality subject to the tyranny of a machine and the wisdom of hands dedicated over the course of a lifetime to construction. There was no room for improvisation, and yet the room was fashioned with intelligence, imagination, and dedication. I learned from those German and Italian manufacturers what we sometimes fail to intuit from schools or books over the course of many years. Our project can be constructed entirely through this process, employing a module whose versatility allows it to be divided or added onto, thus becoming autonomous or dependent on another structure. Our main module, for example, is a kind of loft divided in half in order to create two rooms, as simple as that. In summary, the module is a two-, three-, or four-bedroom apartment; a house can be formed by adding on two or four more modules. The important thing is the versatility of this structure, one that can be entirely factory-made then raised on site in a friendly manner. Miguel Ángel Aragonés

冥想

Medita

一般而言，大多数建筑设计的功能都与成本效益息息相关——成本最低、最简单的方法通常最受欢迎，而且至关重要的是，建筑的使用者在工作时不会被迫放慢速度、产生困惑或注意力分散的状况。但是，纵观建筑史，我们发现该学科逐渐被赋予一种崭新的使命，即，创造一种超脱于日常生活并鼓励冥想和提升精神境界的空间效应。

建筑师的工具包括空间的移动方式、建筑大小和形状以及建筑的结构系统和装饰系统所蕴含的文化内涵。但此外，建筑师还会操控使用者所有的基本感官需求，包括光线的质量、材料的类型和质感、声音及气味，以此带给使用者超乎日常的感官体验。

在本篇文章中，我们将仔细观察一些近期的建筑结构，它们运用了各种不同的方法，旨在为人们创建一个冥想或精神空间。这些设计可能很小，且又是当代建筑，但是，作为建筑方法的一部分，建筑师都采用了不受时间影响的、能展现基本自然效果的设计方式。

Most functions in architecture are a matter of cost and efficiency - the cheapest and most simple method is usually best, and it is vitally important that users of a building are not slowed down, confused or distracted as they go about their tasks. But throughout the history of architecture, the discipline has been called upon to create spatial effects that are beyond the everyday life, and that encourage meditative or spiritual states of mind in the user.

The tools at an architect's disposal include the ways in which space is moved through, the size and shape of the building, and the cultural meanings within its structural and decorative system. But furthermore, the basic senses of the user - the quality of light, the type and texture of material, sounds and smells can all be manipulated by the architect in order to bring the user out of their sense of everyday experience.

In this article we will examine some recent structures that use various strategies to provide meditative and spiritual spaces. The designs may be small, and contemporary, but they all bring timeless manipulations of basic natural effects into play as part of their architectural methodology.

从坟冢到教堂

在全世界范围内，建筑的最初目的通常与精神世界密不可分。许多现存于世的年代久远的建筑都与死亡密切相关——从坟冢到大金字塔，无一不是其鲜明的体现。古代社会耗费巨大的人力、物力建造了一个个用于埋葬尸体的巨大空间，从其功能而言，这些努力都是冗余的，然而很显然，这些空间也是全神贯注于此功能的。

随着时间的推移，庙宇、圣祠以及圣殿与军事建筑和宫殿建筑共同成为最高层次的建筑类型。但是，有别于军事建筑的防御特性和宫殿的富丽堂皇，它们的功能需求更独特，举办的活动更具抽象性和仪式性。宗教建筑不仅必须给人们提供一个可以冥想生存问题的空间，同时在某种程度上也必须是宇宙体系的一种体现。

例如，一座中世纪教堂通常在其布局、结构形式及装饰特征上体现出了对宇宙的一种概念理解，正如基督教徒所理解的那样，同时他们与上帝之间的关系的本质也可通过其建筑设计呈现出来。人们在建筑空间内是如何移动的，建筑空间对人们的各种感官又产生了怎样的影响，这些设计方法都有意识地带领人们脱离日常生活，进入冥想的境界。

这可能是因为我们现在生活在一个不那么神秘的社会里，而且科学现实主义也不需要通过建筑手段来传递信息。但是人体的发展速度远落后于文化，所以在各式各样的当代建筑中，我们依然可以从中看到

From Burial Mounds to Secular Chapels

Across the world, the original purpose of architecture was frequently spiritual. Many of the earliest buildings that still exist are structures related to the dead - from burial mounds to the great pyramids, ancient societies put huge effort into creating grand spaces to entomb bodies, a functionally redundant endeavour, yet one which clearly preoccupied them.

Over time, temples, shrines, and sanctuaries took their place alongside military and palatial structures at the top of the hierarchy of building. But unlike defense or luxury, their functional requirements are strange, the activities they contained highly abstracted and ritualistic. Spiritual architecture must not only provide space for meditation upon existence, but must also be, to an extent, an embodiment of a cosmological system.

For example, a medieval cathedral often expresses - in its layout, structural forms and its decoration - a conceptual understanding of the universe as Christians understood it, and the nature of their relationship with their God can be discerned through the architecture that they created. The ways in which people moved around the space, and the effect it had on their senses, were intended to put them out of the everyday world and into a meditative realm.

It may be the case now that we live in a less mysterious world,

普世教堂，墨西哥
Ecumenical Chapel, Mexico

宗教建筑的影子，即使其中有些建筑根本不含有宗教目的。下面我们将着重研究几个项目，它们运用不同的当代建筑设计方法营造出了冥想的空间效果。

在宗教领域中，身体上的与世隔绝过程通常是精神生活的重要层面。基督教隐士的传统就证实了这一点，剥夺物理和空间上的联系是不断加深对上帝的认识的先决条件，而许多文化都有其牢固的朝圣传统。

在墨西哥的一个周末度假屋的地面上，BNKR建筑师事务所建造了一个小型的普世教堂。作为完全由一个家庭私有的建筑，它不需要看起来像是公共建筑，所以建筑师得以集中精力考虑建筑内部的交通流线设计，从而更好地远离现实世界的喧嚣。该建筑从地面往下下沉了一层的高度，居住者只能沿着一条围绕着教堂的曲线外形而建的下降坡道进入教堂。

由于只能慢慢地向下走，且在很大程度上消除了外部环境的干扰，所以访客才得以远离日常世界，一步步进入光线昏暗、被磨砂玻璃包围的内部空间。而一进入建筑内部，屋顶的圆孔再次带来室外的光线，这是一种产生精神共鸣的姿态，屋顶的水池、围绕这座下沉建筑的绿墙以及教堂中心的一大块石英晶体等简单的设计元素更是强调了这种姿态。

寻求理想的冥想居住空间以达到身体上的与世隔绝，可以通过非

and scientific realism does not require architectural means to communicate its message. But the human body evolves far more slowly than its cultures do, and it is possible to see traces of spiritual architecture in all kinds of contemporary building, even those that have no religious purpose. Below we will examine a few projects that use highly varied contemporary architecture to create meditative spatial effects.

In religious contexts, the process of physical separation has often been a vital aspect of the spiritual life. The Christian tradition of the hermit attests to this, the deprivation of physical and spatial contact as a precondition for a heightened awareness of God, while many cultures have strong traditions of pilgrimage.

In the grounds of a weekend residence in Mexico, BNKR Arquitectura have built a small ecumenical chapel. As a fully private building for one family, it need not present a civic presence, so the architects have managed to concentrate their energies on the journey through the building, accentuating its removal from the everyday world. The building is sunk a full storey height under the ground level, and the inhabitant can only enter by circling the building via a descending ramp that follows the curved outline of the chapel.

By being forced to approach slowly, and by largely removing the external context, the visitor is distanced from the everyday world until they enter the dimmed inner space, which itself is surrounded by frosted glass. Once inside, however, an oculus in the ceiling re-presents the light of the exterior in a circular frame, a spiritually resonant gesture, one that is accentuated by simple elements - a pond on the roof, green walls

圣雅各布小教堂，德国菲施巴豪
St. Jacob's Chapel, Fischbachau, Germany

常简单的方式实现。在德国巴伐利亚乡村，建筑师米歇尔·德·卢基建造了一座可俯瞰全村的小教堂。该建筑遵循当地教堂建筑的传统，设计了一个入口门户，通向狭小而黑暗的内部空间。只有在黑暗的环境中上几步台阶，来访者方可到达长椅的位置，长椅正前方是一扇圆形的窗户，窗户外面有一个木制十字架。从简洁而传统的设计、粗糙的混凝土砖块和简单的木结构细部设计来看，这座新教堂有一些阿尔多·罗西早期作品的忧郁风格。与附近不知名的墨西哥教堂相比，尽管这座教堂很小巧，但很有地位。许多有宗教目的的建筑都依赖于其偏僻的地理位置和简洁的设计来创建一个用于冥想的安静空间。但这些品质往往是难以获得的，而且在心理上与尘世喧嚣创造距离更是难以实现。

就这一点而言，日本东京的Ekouin Nenbutsudo神寺面临两大挑战。首先它是一座城市建筑，坐落于世界上最大城市之一的街道上；其次它不仅要为祈祷者和集会活动提供合适的空间，也要提供住宿和培训设施。尽管存在种种限制，该建筑也必须给前来寻求宗教体验的人提供一个朝圣之所。

目前，建筑师Yutuka Kawahara运用多种方法解决了Ekouin Nenbutsudo神寺所面临的难题。其中一个主要措施就是将建筑物的地板结构延伸至外围护结构之外，创造了一些露台，强度足以支撑土壤层，然后栽满竹子。这些绿色植物主要用于减少来自附近街道的污染以及避开喧嚣。同时，当花园景致过滤了来自外部的视线时，建筑能呈

surrounding the sunken building, and a large quartz crystal at the center of the chapel itself.

The process of physical estrangement to provide appropriate space for meditative inhabitation can be achieved with remarkably simple means. In the Bavarian countryside of Germany, Michele de Lucchi built a small chapel that looks out over the local fields. The building, which follows on from a local tradition of small chapel buildings, provides a portal doorway that leads into a very small and dark space. The visitor must rise a few steps in the gloom to reach a bench, which is positioned right behind a circular window, presenting a view out to a wooden cross that has been set out in the field beyond. This new chapel has something of the melancholy quality of the early work of Aldo Rossi, with its stripped back but clearly classical sensibility, and its rough concrete blocks and simple timber detailing. Compared to the near invisibility of the Mexican chapel, it is monumental, despite its tiny size. Many buildings with spiritual purposes rely both on their physical remoteness and also on their simplicity of programme to create the setting of calm required for their meditative purposes. But these qualities are often unavailable, and the creation of the required mental distance from the world is far harder to achieve.

The Ekouin Nenbutsudo Temple in Tokyo, has two challenges in this regard. It is first of all an urban building, taking its place on a street in one of the world's largest cities, and secondly it must provide not only spaces for prayers and congregations, but also the facilities for accommodation and training. Yet despite these restrictions, the building has to provide a sanctuary for religious experience.

照片提供：©Naomichi Sode

东京Ekouin Nenbutsudo神寺，日本
Ekouin Nenbutsudo Temple Tokyo, Japan

现出一种宁静之感。

光的抽象概念也是该建筑的一个重要组成部分，好似一根根链条从建筑拱腹边缘奔流而下，这个位置悬挂了许多多面玻璃串。这些悬挂式水晶饰物与周围的氛围交相呼应，同时也印证了它们是在不断运动的事实：反射和折射光线，向建筑内部空间投射各种闪烁的光线和色彩。但是这些效果就一定与宗教有关吗？20世纪，一些最享有盛名的宗教建筑都是由与客户的宗教信仰没有任何联系的设计师创作的。最著名的案例就是在第二次世界大战结束后的那些年里，天主教会邀请了许多现代主义建筑师来设计教堂，在勒·柯布西耶的朗香教堂这类作品中，建筑师开发了一系列空间效果，与礼拜仪式或礼拜方式相比，这些效果似乎与存在主义艺术和文学作品有更多的相通点。通过接触与灵性有关的项目，建筑师逐渐掌握了全新的空间效果设计方法。

众所周知，美国的大学都会有遗赠和捐赠活动，这些资金通常都被投入到了建筑创作中。位于加利福尼亚的斯坦福大学委托Aidlin Darling设计公司创作一个非宗教空间——茶隼冥想中心，在这里，学生和教职员工都可以暂时逃避现实生活的压力，同时也可远离忙碌的精英学术生活。因此该建筑涵盖了教堂、画廊和花园凉亭的功能，并分享了许多前面描述的创造宗教空间的技术。

该建筑建在一个基座上，地面和天花板都采用了一种很黑的染色木材，而Nathan Oliviera创作的艺术品挂于空间内。建筑物的周围有一

Architect Yutuka Kawahara has approached this challenge in a number of ways. One of the prime gestures has been to extend the structural floor out past the envelope of the building to create terraces strong enough to support earth beds, which have then been planted with bamboo. This planting is designed to reduce pollution from the nearby streets, as well as muffle noise. It also serves to give the building a tranquil feeling, when all external views are filtered through the sight of a garden.

The abstraction of light plays a role in the building here as well, as chains run down from the edge of the soffits on which faceted glass is strung. The interaction of these hanging crystals with the atmosphere means that they are in constant motion, reflecting and refracting light, casting glints, beams and spectra into the inner spaces of the building. But are these effects necessarily religious? Some of the most famous religious architecture of the 20th century was created by designers with no connection to the faith groups that were their clients. The most famous example was the Catholic Church's patronage of modernist architects in the years following WWII, and in works like Le Corbusier's Ronchamp chapel, a whole series of spatial effects were developed that seemed to have more in common with existential art and literature than the liturgy or modes of worship. Through commissions that necessitated engaging with spirituality, new spatial effects passed into the architect's repertoire.

American universities are well known for bequeathments and donations, which often go into the creation of architecture. Stanford University in California commissioned Aidlin Darling Design to create the Windover Contemplative Center, a non-

斯坦福大学茶隼冥想中心，美国
Windhover Contemplative Center, USA

些花园，花园里有许多栽种在粗砾石中的植物和一个巨大的水池。

该建筑的主要基本构成元素是一些划分空间的夯土墙，无论从它们与地面和天花板稀少的关联，还是从其斑驳、起伏不平的外表来看，这些墙都让人联想到了密斯·凡·德·罗在两次世界大战期间的开创性作品中的缟玛瑙墙。夯土中的土壤是精心挑选的，与艺术品形成了互补的作用，创造的视觉效果与宗教建筑的基本品质大同小异，但其操作过程却是截然不同的——建筑师没有以抽象的方式通过将感官分离来纯化体验，而是使这些墙面保留"与土地有关"的厚重的特点，是地质的抽象而不是宇宙的抽象。

即使当宗教建筑中一部分鲜明的宗教因素被移除时，建筑效果往往还是对自然进行调查研究的体现，或是表达了与大自然的重要关系。宇宙秩序的抽象概念及模式总会在建筑设计中有所显现，尤其以宗教建筑最为明显。

religious space where students and staff can visit to escape the stress and bustle of elite academic life. The resulting building is part chapel, part gallery, part garden pavilion, and shares many of the techniques of creating religious space described previously.

The building is raised on a plinth, and both floor and ceiling are created from a very dark stained timber, while artworks by Nathan Oliviera are hung around the space. Surrounding the building are gardens that include plants set in gravel and a large reflecting pond.

The main primal, elemental part of the building are a number of rammed earth walls that divide up the space, reminiscent of the onyx walls from Mies van der Rohe's seminal inter-war work, both in terms of their slightly unconnected relation to the floor and ceiling planes, but also their mottled, rippling appearance. The rammed earth, with soil chosen carefully to complement the artworks, creates an effect that is similar to the elemental qualities of the religious buildings, but operates in almost exactly the opposite direction - instead of rarefying the experience through splitting senses off in an abstracted manner, the walls present themselves as heavy and "grounded", a geological abstraction rather than cosmic.

Even when the explicitly religious parts of a spiritual building are removed, the architectural effects so often amount to an investigation into, or an emphasised relationship with, nature. Abstractions and models of cosmic order are almost always present in architecture, but are especially so in religious buildings. Douglas Murphy

斯坦福大学茶集冥想中心

Aidlin Darling Design

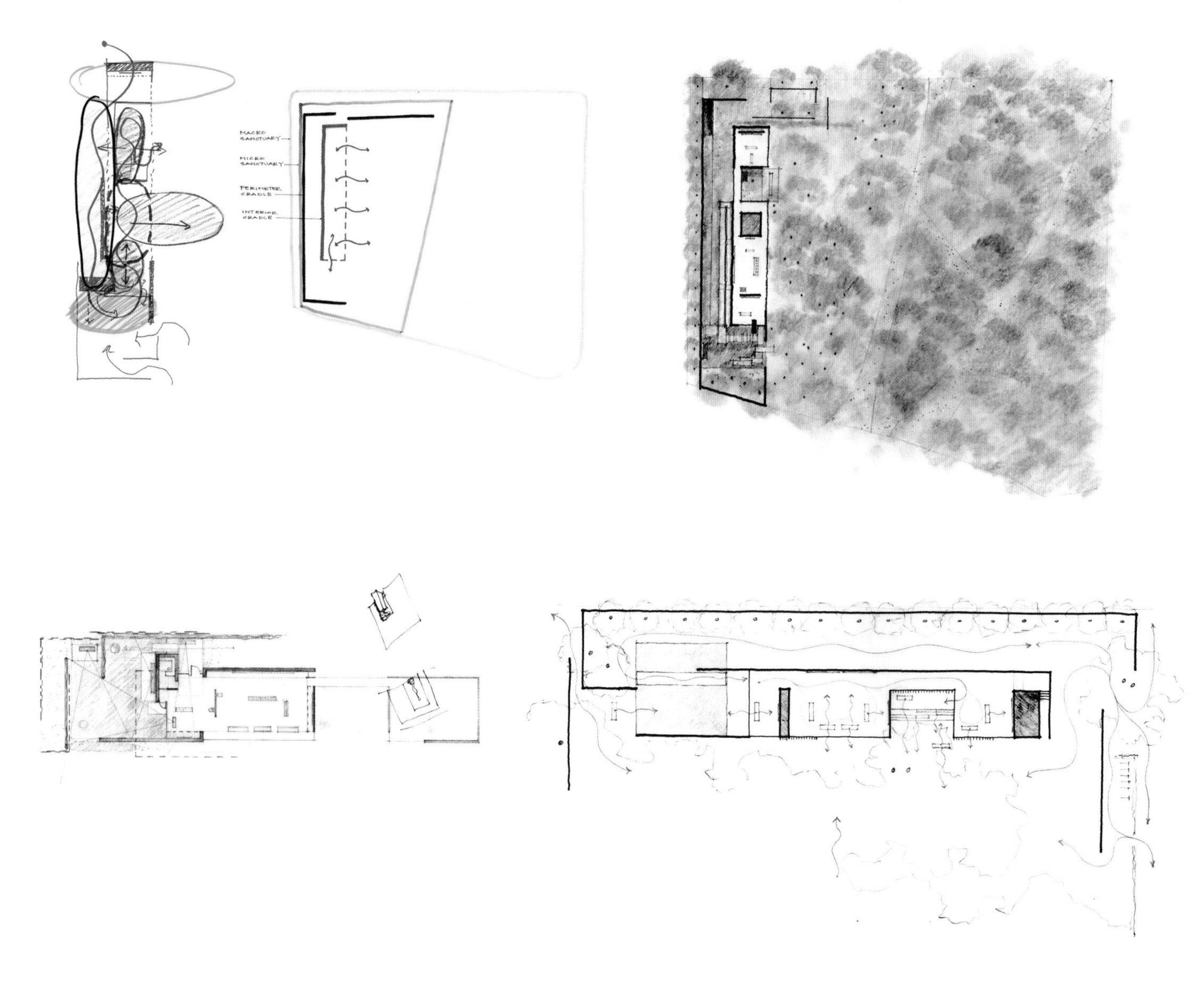
MACRO SANCTUARY
MICRO SANCTUARY
PERIMETRE CRADLE
INTERIOR CRADLE

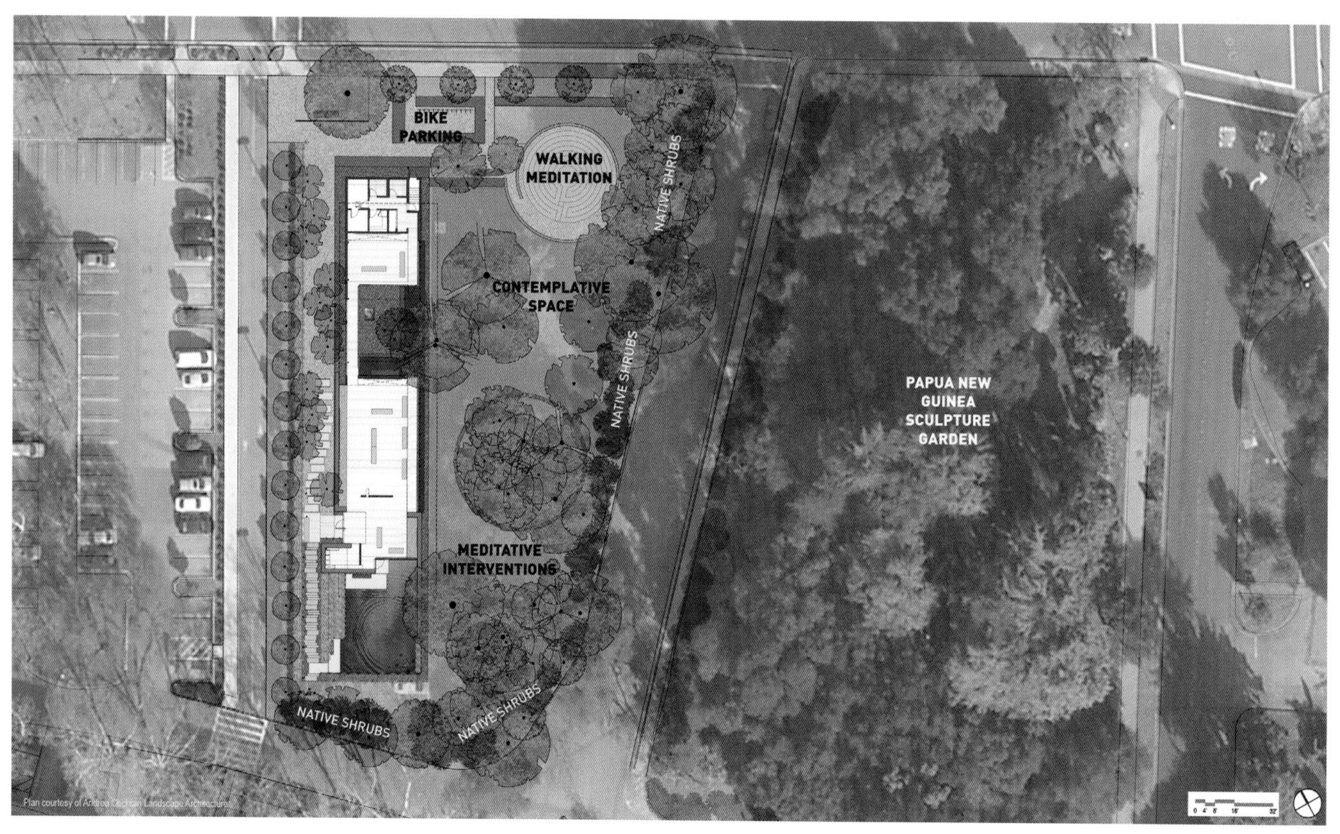
BIKE PARKING
WALKING MEDITATION
NATIVE SHRUBS
CONTEMPLATIVE SPACE
NATIVE SHRUBS
PAPUA NEW GUINEA SCULPTURE GARDEN
MEDITATIVE INTERVENTIONS
NATIVE SHRUBS
NATIVE SHRUBS

每个人为了保持存在感都需要去滋润他的精神世界。这组关于茶隼的画就是为了帮助人们去关注自身——让人格魅力展翅翱翔!

——内森·奥利弗瑞（1928—2010）

奥利弗瑞创作了“茶隼”系列画作，灵感源于一只由斯坦福山麓俯冲而下的茶隼，是为了唤起人们暂时脱离日常生活、感受飞翔快意的情感；他生前希望该系列绘画可以放在一起，放在一个供人冥想的空间内展示。为了帮助大学生释放紧张情绪、自我调节世界观和人生观，斯坦福大学希望建立一个非宗教性质的冥想空间，该空间内将展示内森·奥利弗瑞的“茶隼”系列画作，给人以启发，同时也是“茶隼”系列绘画的展示厅。为避免制造出类似博物馆似的氛围，学校旨在建造一个相对更加有机的空间，它不仅以某种方式与自然相融，同时也为师生躲避紧张的日常生活提供了一个“避难所”。这个用于安静思考和反思、沉思的独特冥想空间对斯坦福大学的学生、教职员工乃至整个社区的成员全天开放。

茶隼冥想中心集艺术、景观和建筑为一体，为使用者建立了一个重振精神、鼓舞士气的空间。该中心位于整座大学校园的中心，毗邻一片天然橡树林，以前是停车场。到达建筑的入口需要穿过由一排高耸的青葱竹林环绕的长长的私密花园，这样斯坦福校区的访客在进入建筑之前就可以摆脱外界的纷杂喧嚣。在建筑内部，该空间完全对东侧的橡树林以及远处的巴布亚新几内亚雕塑公园开放。

透过装有百叶窗的采光天窗洒下来的自然光线照亮了奥利弗瑞的那些4.6m到9.1m长的不朽画作，其余空间则有意识地保持黑暗，让访客更加专注地欣赏自然光线下的画作及远处的美丽景观。厚实的夯土墙和木贴面进一步从听觉、触觉、嗅觉和视觉方面提升了访问者的感官体验。

贯穿于整座建筑的水与景观相结合，有助于静心冥想；主展厅和庭院中的喷泉负责营造环境声效，而南侧的水池静静地倒映着周围的绿树。外部的冥想空间也与中心的功能相结合，来访者既可以安静地欣赏画作，又可以饱览自然景观。来访者不用靠近建筑物，从东侧的橡树林就能欣赏到中心内的画作，该设计有效地为整个斯坦福大学提供了一座全天开放的精神庇护所。

Windhover Contemplative Center

"One must nurture the spirit identity within one's self in order to fully exist. The Windhover paintings are intended to assist people in centering themselves – and to allow the human spirit to fly."

– Nathan Oliveira (1928—2010)

Oliveira created the Windhover series, inspired by the kestrels swooping above the Stanford foothills, to evoke the feeling of flight and detachment from the everyday; it was his wish that the paintings be displayed together in a place set aside for contemplation. Recognizing the need on campus for a space for students to relieve stress and gain a greater perspective of one's life and the world in general, Stanford University sought to create a non-denominational space for contemplation, a space that would be inspired by and display paintings from Oliveira's Windhover series. The university wanted to avoid creating a museum-like environment in search of a more organic space that engaged nature in some manner while providing a refuge from the intensity of daily life. The Contemplation Center is intended for quiet reflection throughout the

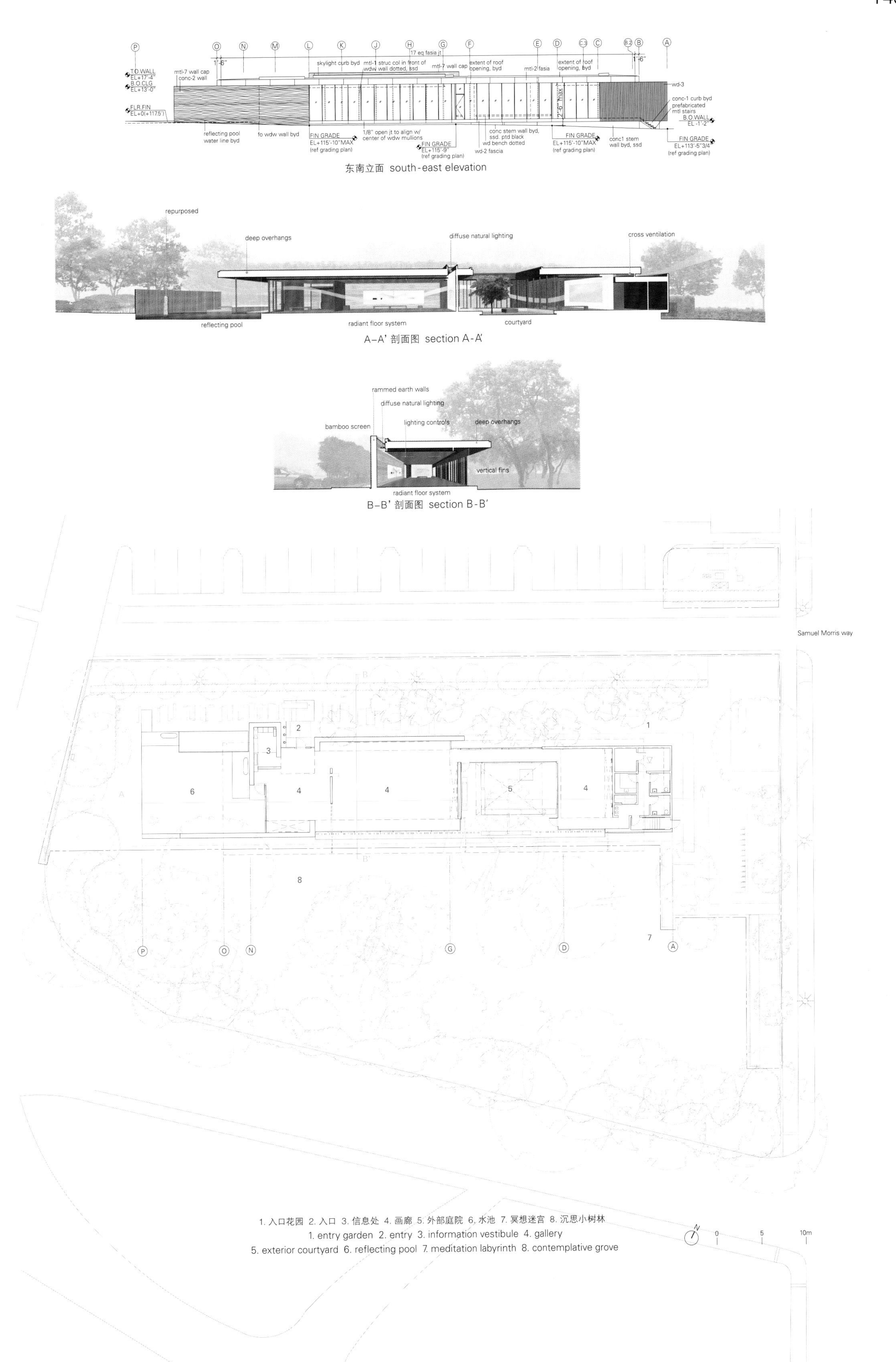

东南立面 south-east elevation

A–A' 剖面图 section A-A'

B–B' 剖面图 section B-B'

1. 入口花园 2. 入口 3. 信息处 4. 画廊 5. 外部庭院 6. 水池 7. 冥想迷宫 8. 沉思小树林

1. entry garden 2. entry 3. information vestibule 4. gallery
5. exterior courtyard 6. reflecting pool 7. meditation labyrinth 8. contemplative grove

项目名称：Windhover Contemplative Center
地点：Stanford University, 450 Serra Mall, Stanford, CA 94305, USA
建筑师：Aidlin Darling Design
项目建筑师：Kent Chiang
项目经理：Roslyn Cole
项目团队：Melinda Turner, Michael Pierry, Jeff LaBoskey
结构与土工工程师：Rutherford & Chekene / 土木工程师：BKF Engineers
机械暖通工程：Air Systems, Inc. / 电气工程：Elcor Electric
景观设计师：Andrea Cochran Landscape Architecture
照明设计师：Auerbach Glasow French
自然采光顾问：Loisos + Ubbelohde
声学顾问：Charles M. Salter Associates, Inc.
总承包商：SC Builders, Inc.
夯土承包商：Rammed Earth Works
用地面积：371.61m² / 设计时间：2014
摄影师：©Matthew Millman (courtesy of the architect)

day for any Stanford student, faculty, or staff member, as well as for members of the larger community.

The Windhover Contemplative Center is conceived as a unification of art, landscape and architecture to both replenish and invigorate the spirit. The sanctuary is located in the heart of the campus on a former parking lot adjacent to a natural oak grove. The extended progression to the building's entry through a long private garden, sheltered from its surroundings by a line of tall bamboo, allows members of the Stanford community to shed the outside world before entering the

sanctuary. Within, the space opens fully to the oak grove to the east and the Papua New Guinea Sculpture Garden beyond.

Louvered skylights wash the monumental 15 to 30 foot long paintings in natural light. The remaining space is kept intentionally dark to focus the visitor's attention on the naturally highlighted paintings and the landscape beyond. Thick rammed earth walls and wood surfaces further heighten the visitor's sensory experience acoustically, tactilely, olfactorily, as well as visually.

Water, in conjunction with landscape, is used throughout as an aid for contemplation; fountains within the main gallery and the courtyard provide ambient sound while a still reflecting pool to the south reflects the surrounding trees. Exterior contemplation spaces are integrated into the use of the center, allowing views to the natural surroundings as well as to the paintings within. From the oak grove to the east, visitors can view the paintings glowing within the center without accessing the building, effectively creating a sanctuary for the Stanford community day and night.

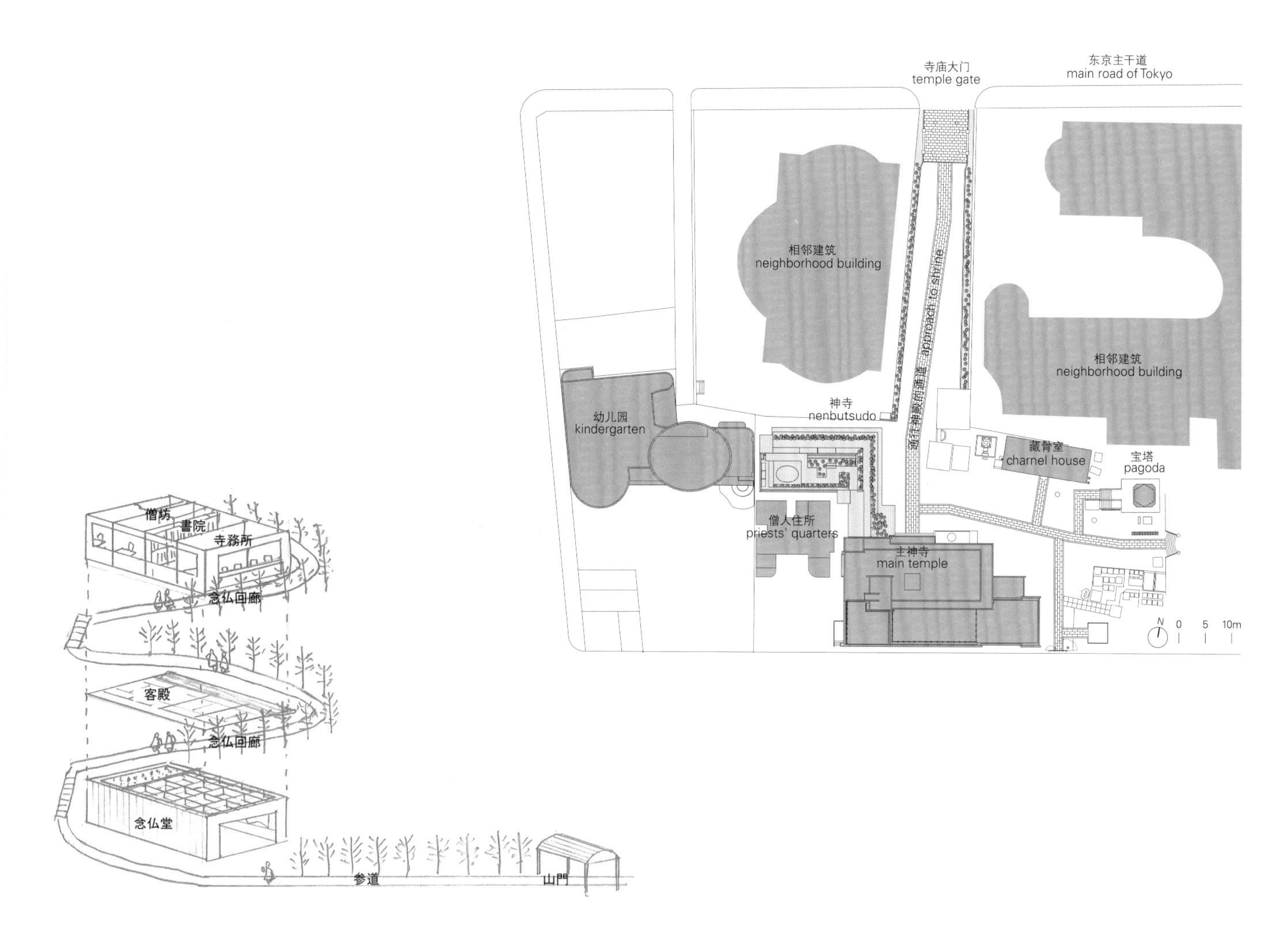

东京Ekouin Nenbutsudo神寺

Yutaka Kawahara Design Studio

该项目是佛寺的替代品，现坐落于日本东京的市中心。与佛寺相同，这座神寺主要有三大功能。由于空间有限，无法同时在一个楼层设计三个功能，因此，我们决定设计一座多层建筑。为了将三大功能联系起来，我们在建筑外部设计了一条通向主神寺的小路。

每个功能都有其各自的一套要求。例如，一层是厚实的混凝土墙和分隔层，主要用于隔声和防火防盗。与此相反，为了便于沟通互动，二层主要由钢框架构成。三层的空间经过精心设计，专用于举办各种较为私人的活动。这三层楼在建筑内部并没有衔接，但都通过栽种了竹子的室外走廊紧密相连。来访者必须通过这些走廊才能到达每个功能楼层。三大功能在空间上的联系与竹子小道并不是独立存在的。这可以从过道的设计中看出来，这种设计对场地产生了一种立体化的扩展效果。

该项目的主要目标是在这个大都会的市中心营造一个巨大的绿洲。为了与现有的景观和谐相融，在这里种植了竹子。竹子的存在并不会对道路产生干扰，相反，它在喧嚣的市中心打造了一个绿色空间。其中也包括由108块施华洛世奇水晶制成的玻璃竹子。为了与主神寺中的佛像数量保持一致，建筑师采用施华洛世奇雕刻了40座佛像雕塑。在建筑设计中使用施华洛世奇作为外部建筑材料是很罕见的。玻璃竹子一经制作完成，便在工厂里接受了反复检验，最终靠拉力将其固定；它跟用于佛教祈祷的念珠很相似，而且也像棱镜一样闪闪发光，象征着进入天堂或涅槃的人们。

Ekouin Nenbutsudo Temple Tokyo

This project is the replacement of a Buddhist temple in the center of Tokyo, Japan. The temple will have three functions, consistent with Buddhism. However, the site had no longer enough space on one level for all three functions, so we decided to vertically stack them. In order to link the functions, an outside walkway to the main temple is planned.

務所入口

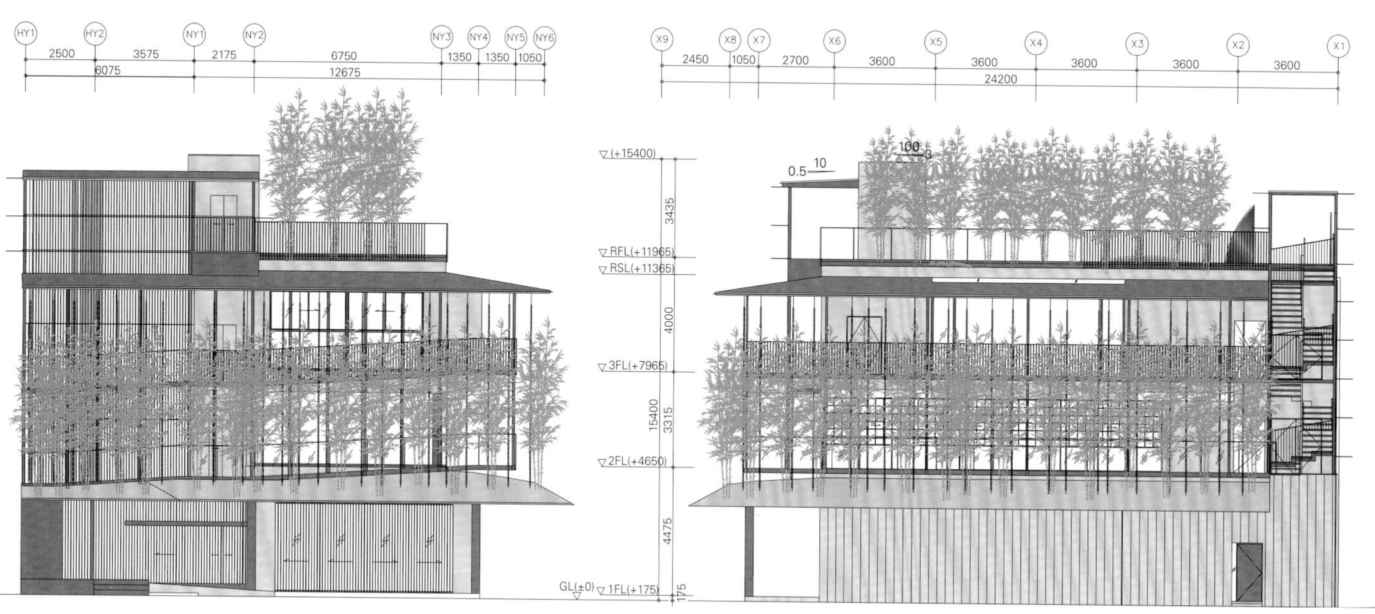

东立面 east elevation

北立面 north elevation

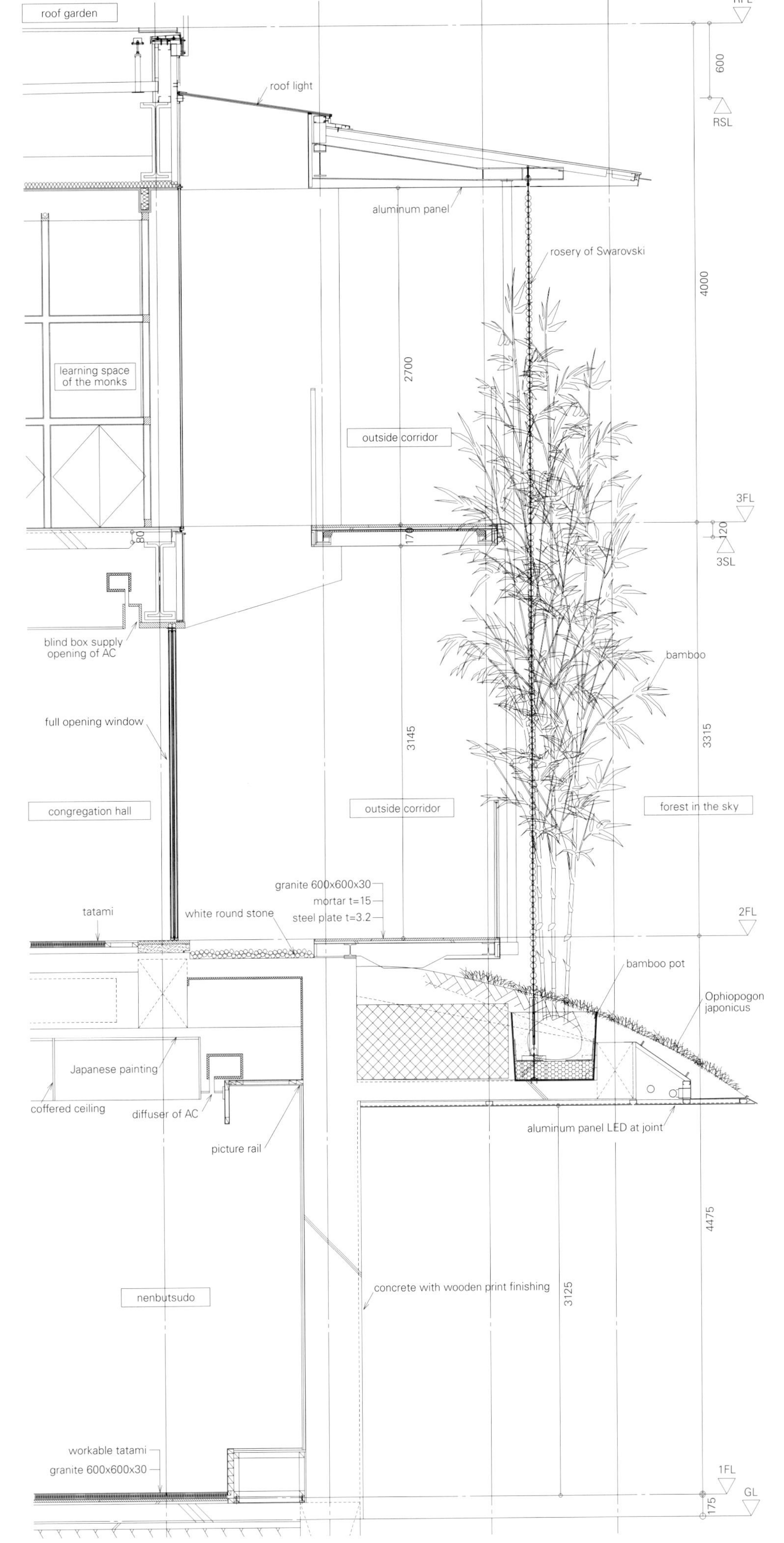

A–A' 剖面详图 detail section A-A'

Each function has its own set of requirements. For instance, the first floor has thick concrete walls and insulation for soundproofing, crime prevention and fire prevention. In contrast, the second floor has steel frames which are conducive to interaction. The third floor has spaces that are carefully designed for more intimate activities. These are not connected inside the building, but are linked by the bamboo-covered outside corridor. Visitors have to pass through the corridor to access each function. The vertical connection does not work independently with the corridor. This can be seen from the pathway, and becomes a three-dimensional extension of the site.

The main goal of the project is to create a large oasis in a metropolitan area. In order to harmonize with the existing scenery, additional bamboo was planted. The bamboo does not interfere with the walkway; it creates a green space in

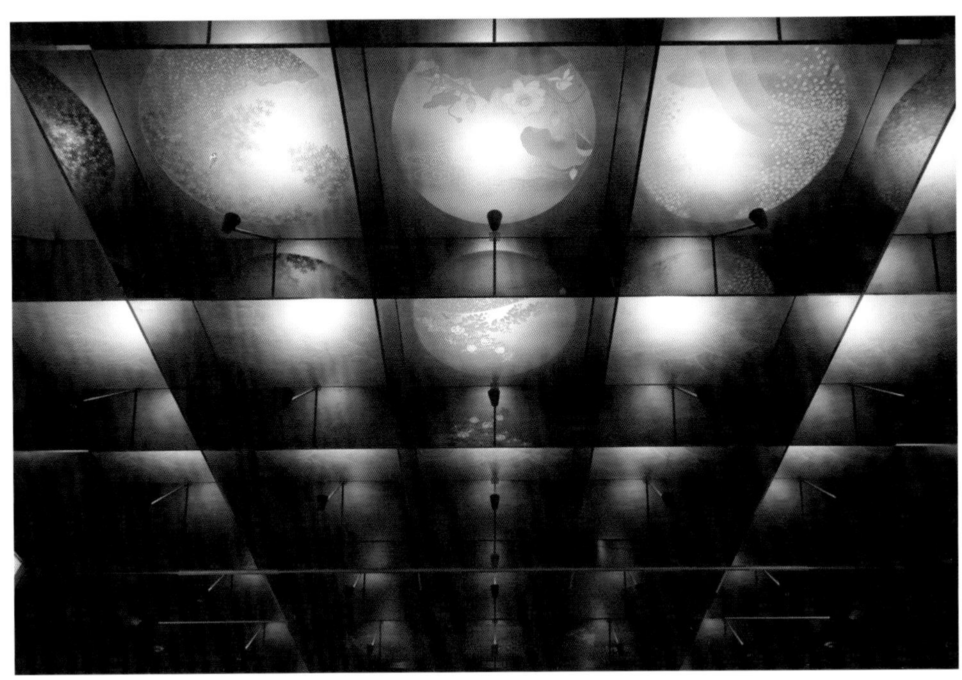

the busy city center. Glass bamboo made of 108 pieces of Swarovski was added to it. Forty statues of Buddha were carved of Swarovski to match the number of statues of Buddha in the main temple. The use of Swarovski as external building material is quite rare in architecture. When the glass bamboo was produced, it was repeatedly examined in the factories, and in the end it was fixed by tension. This is similar to a rosary, which is commonly used in Buddhist prayers; it also glitters like a prism and represents people who enter the world of the "Gokuraku (paradise or nirvana)".

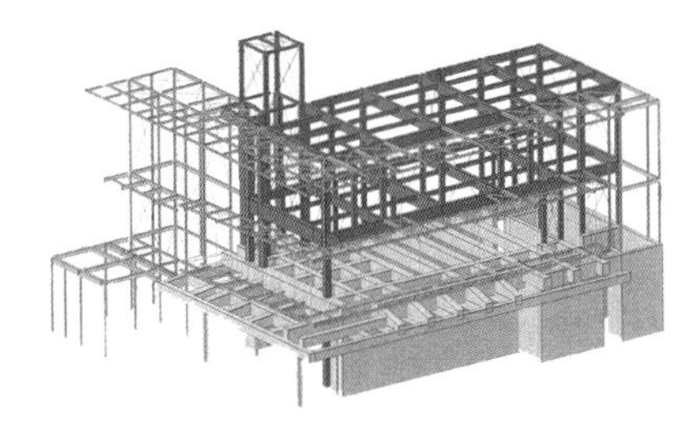

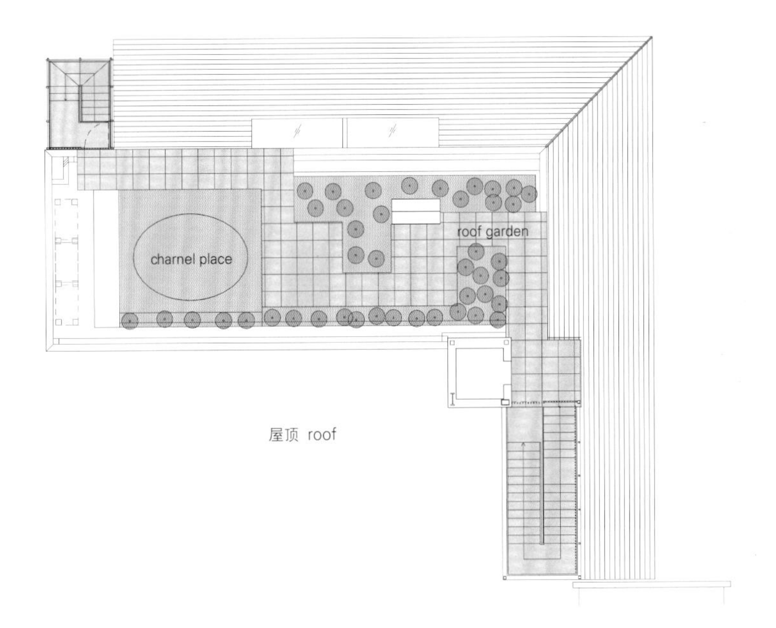

项目名称：Ekouin Nenbutsudo Temple Tokyo
地点：Tokyo, Japan
建筑师：Yutaka Kawahara _ Yutaka Kawahara Design Studio
结构工程师：Satoshi Okamura(kap)
景观设计：Yutaka Tajima _ Landscape design inc
客户：Ekouin / 用途：temple
施工：Matsui Construction co., ltd.
用地面积：1,174.89m² / 建筑面积：300.23m²(others:953.66m²)
总建筑面积：496.69m²(others:2,138.30m²)
景观面积：business district
建筑覆盖率：32.25% / 总建筑覆盖率：67.76%
结构设计：reinforced concrete with steel frame
外饰面：concrete with wooden print finishing
内饰面：tatami, fusuma
设计时间：2011.1—2012.7 / 施工时间：2012.7—2013.3
摄影师：
©Makoto Yoshida (courtesy of the architect) - p.149, p.150~151, p.154
©Naomichi Sode (courtesy of the architect) - p.152, p.153, p.155, p.156~157, p.158

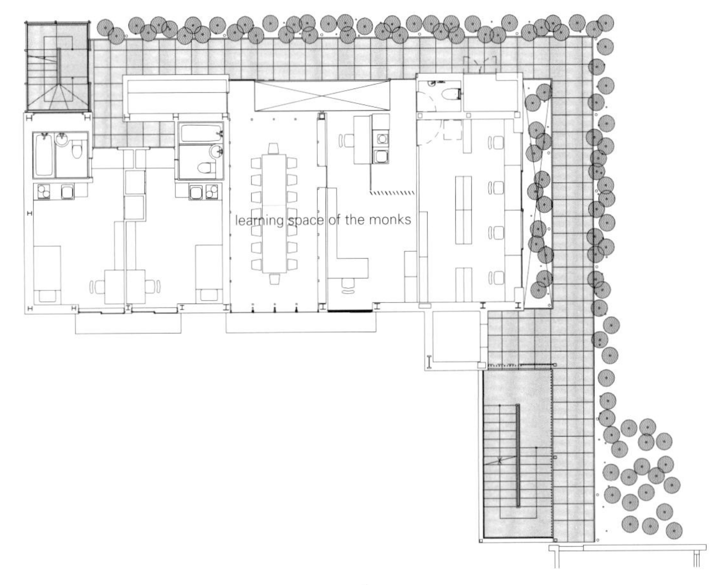

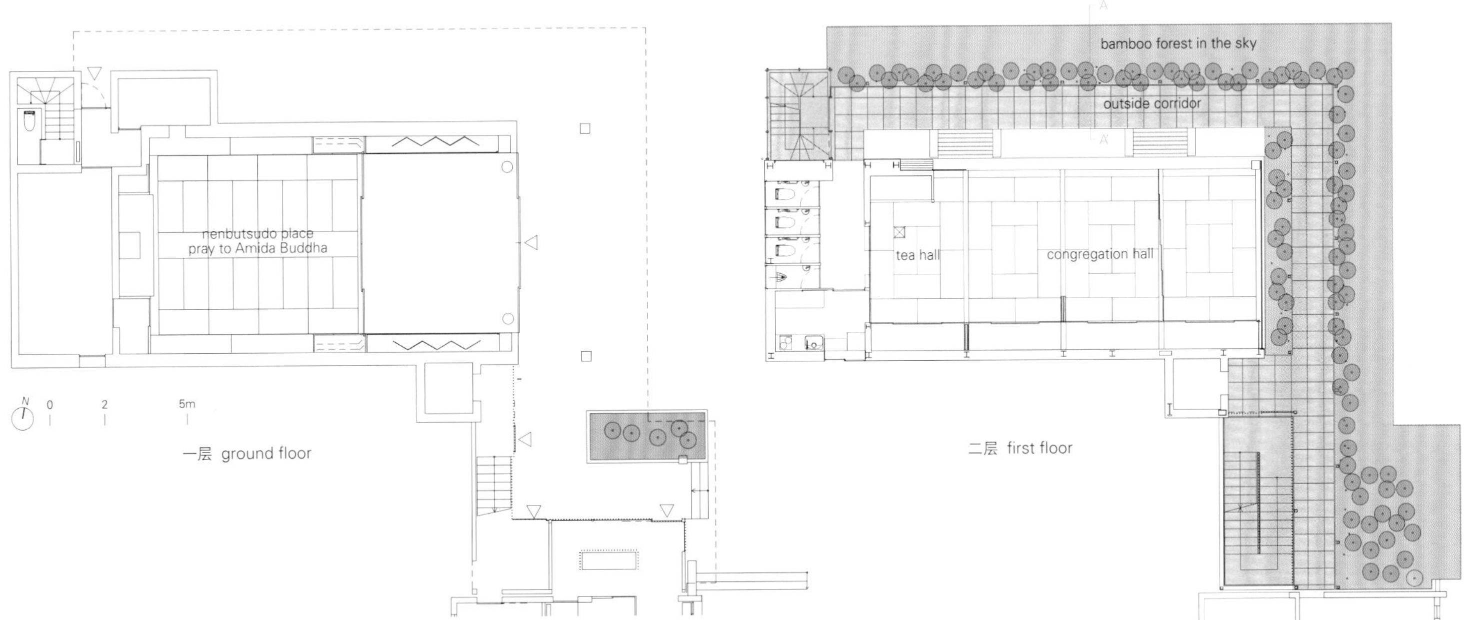

普世教堂

BNKR Arquitectura

客户在看过我们设计的大牧场教堂之后，便委托我们在其位于墨西哥库埃纳瓦卡的周末度假别墅身后建造一个私人小教堂，他们想要一个非宗教性质的通用空间，主要用于冥想。他们很喜欢大牧场教堂，不过他们想要一个设计更加谨慎的、可隐藏于其住宅之中的空间。我们设计的第一个教堂——大牧场教堂是用于举办婚礼的，是对生命的颂歌；第二个教堂——日落教堂则是一个地下墓室，是对死亡的哀悼；这第三座教堂，就是生（大牧场教堂）与死（日落教堂）中间的平衡点，也是我们深入自省的绝佳机会。

这座教堂几乎完全隐藏于地下，访客可沿螺旋通道盘旋进入。斜坡两侧都是植物墙，形成了一个垂直花园。

教堂的房顶由水池构成，水池的中间有一个圆孔，实际上它是金属板上一个镶嵌了玻璃的开口，这样一来，阳光就可以穿过水面在建筑内部产生各种光影效果。这个空间的外侧是由众多独立的玻璃条构成的格栅墙，便于阳光穿透进入内部。屋顶的圆孔在视觉上将内部空间与外部的植物墙和蓝天相连，同时该设计也在不断地提醒我们，周围的一切都与我们紧密相连。教堂中心有一座结构独特的金属喷泉，其上是一块巨大的石英石，主要用于反射透过圆孔的光线。

项目名称：Ecumenical Chapel
地点：Cuernavaca, Morelos, Mexico
建筑师：BNKR Arquitectura
主管合伙人：Esteban Suarez
项目团队：Emelio Barjau, Jaime Sol, Jorge Alcantar, Christian Morales, Gloria Castillo, Montserrat Escobar, Marcell Ibarrola, Fernando Maya, Marco Mayote, Daniel Aguilar
结构工程师：Juan Felipe Heredia
机电工程师：Sylvia Roman
照明工程师：Ricardo Noriegga, Santiago Bautista
建筑面积：170m^2
设计时间：2012 / 竣工时间：2013
摄影师：©Jaime Navarro (courtesy of the architect)

Ecumenical Chapel

The clients who once visited our Estancia Chapel contacted us to design a private chapel in the backside of their weekend house in Cuernavaca, Mexico. They wanted an Ecumenical Chapel, a non-religious and universal space, to meditate. They liked our Estancia Chapel but wanted something more discreet, something that would be hidden from their house. Our first chapel "Estancia" is for a wedding to celebrate life and the second "Sunset Chapel" is a mausoleum in a garden of crypts for mourning death. This third chapel meant for meditation represents the midpoint between these two opposites, life and death, so it is an opportunity to journey into our deeper self.

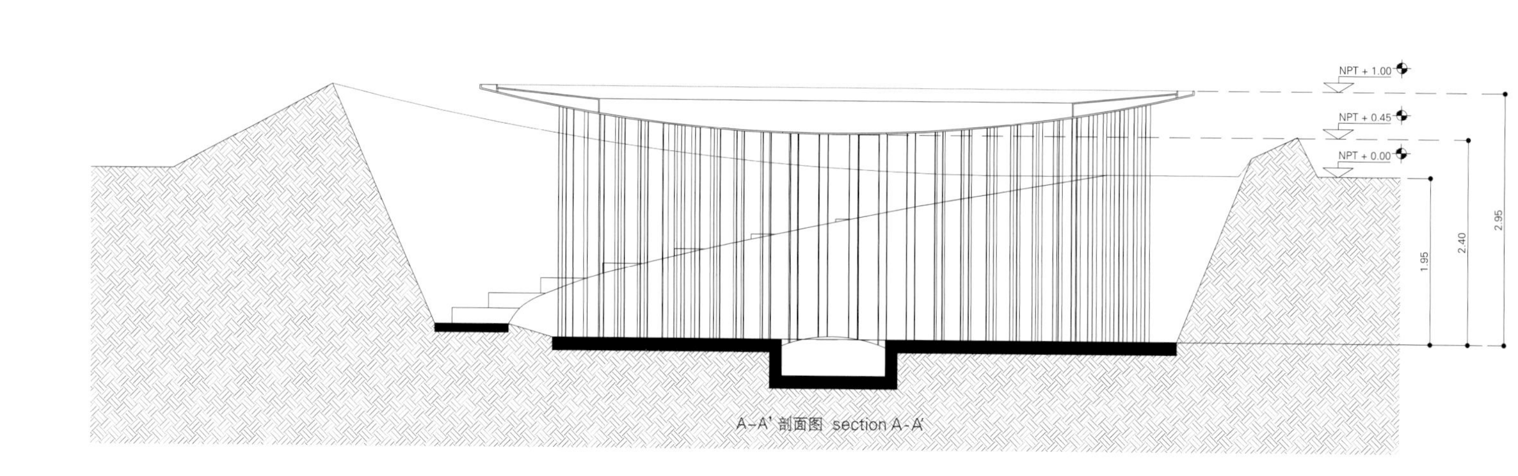

A-A' 剖面图 section A-A'

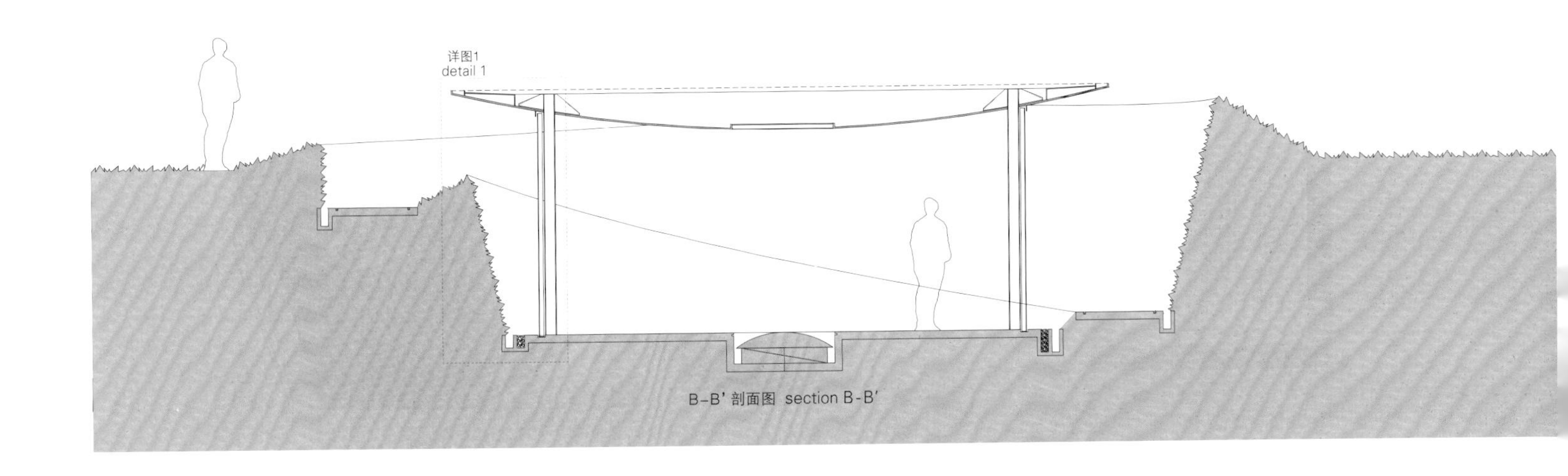

B-B' 剖面图 section B-B'

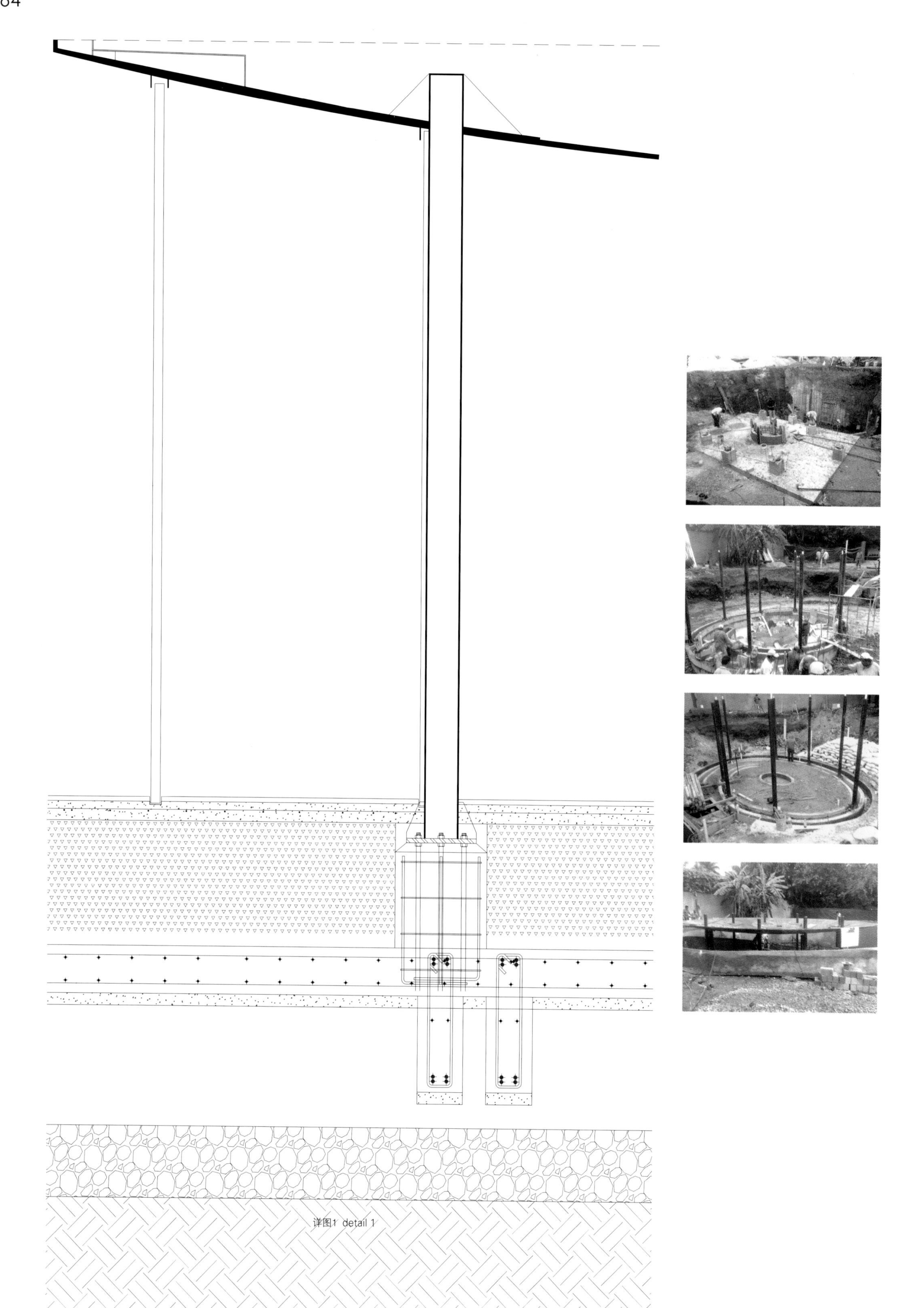

详图1 detail 1

The chapel is buried underground and a spiraling ramp that surrounds it brings us inside. This ramp is flanked with a vegetated wall that functions as a vertical garden.

A water pond forms the rooftop of the chapel and at its center we find an oculus, a glass covered opening in the metallic plate, that lets sunlight filter through the water generating light and shadow patterns on the inside. The space is contained by a lattice wall formed by separated glass beams that lets the air flow through its inside. The oculus is also a visual connection with the outside vegetation and the sky, a constant reminder that we are part of a whole and connected with everything that surrounds us. In the center of the chapel we find a metallic fountain with a giant quartz on top that reflects the light coming through the oculus. BNKR Arquitectura

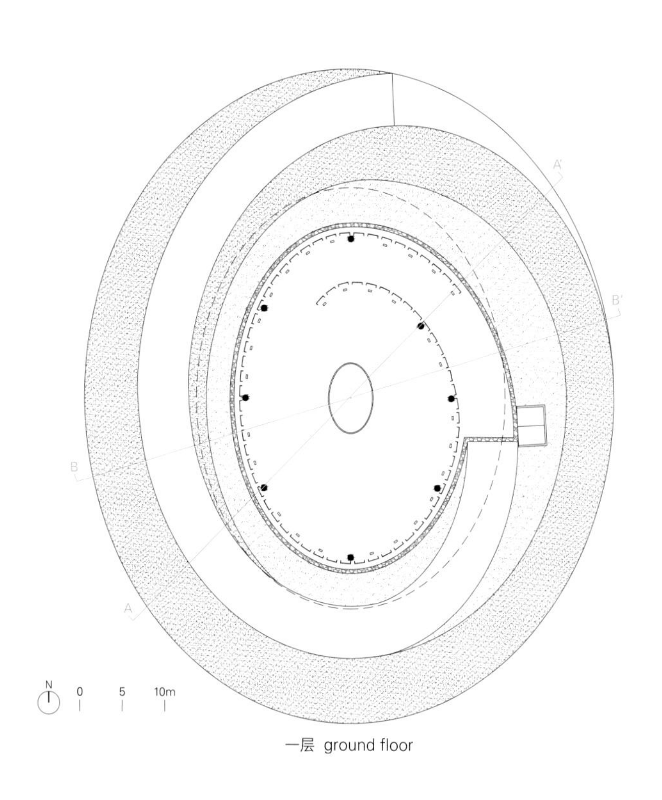

一层 ground floor

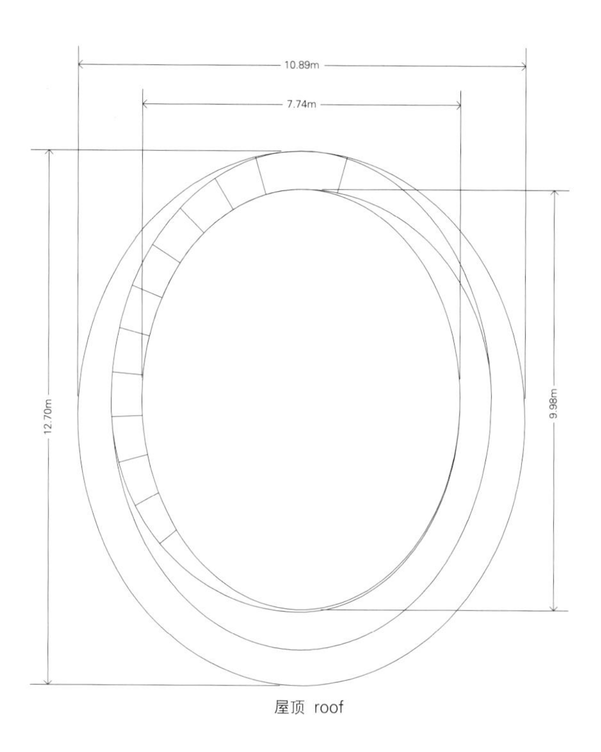

屋顶 roof

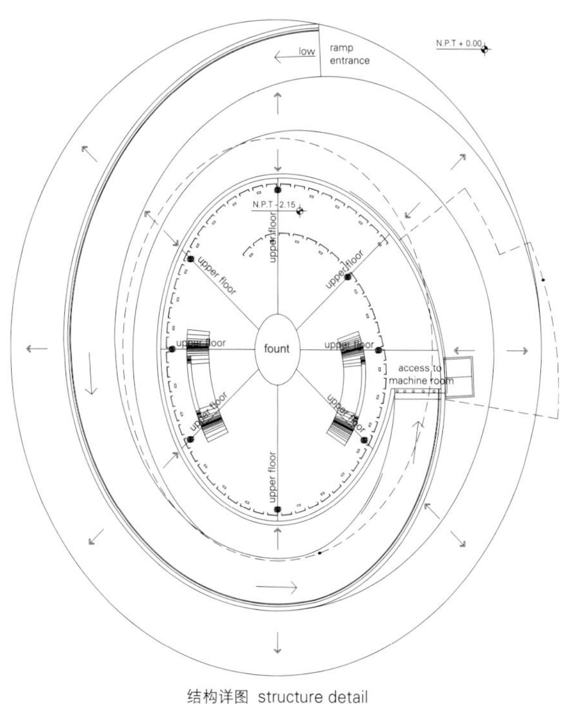

结构详图 structure detail

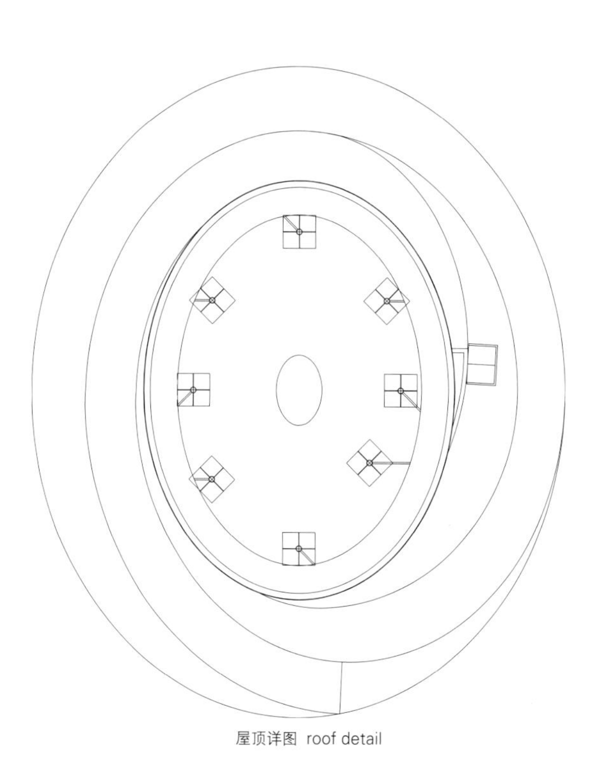
屋顶详图 roof detail

圣雅各布小教堂

Architetto Michele De Lucchi S.r.l.

这座为纪念圣雅各布而修建的还愿小教堂坐落于慕尼黑南部的巴伐利亚州乡下，它位于一个私人地块内，这里是跨越欧洲大陆前往西班牙西北部的圣地亚哥-德孔波斯特拉的至圣所朝拜的必经之路，距离至圣所2347km。该建筑长5m、宽3m，采用了在德国菲施巴豪地区到处可见的神殿的尺寸。建筑内部空荡无物，冥想的设施都在公园外。沿着楼梯走上去，可以看到一条木凳放在圆形窗户前。从窗户望去，可以看到远处的地平线和正立在草坪对面边缘位置的十字架。木凳、大理石的圣水器和木质的烛台架是屋内仅有的家具。当地人都称这座建筑为"上帝的混凝土"，它的屋顶和门窗都是用铜和天然木材制成的。

St. Jacob's Chapel

The small votive chapel dedicated to Saint Jakob is situated in the Bavarian countryside south of Munich, in a private property located along one of the ways across Europe to the sanctuary of Santiago di Compostela at a distance of 2,347 kilometers, in northwest Spain. The building takes up the measurements (5 x 3 meters) of the numerous religious

shrines to be found in the surrounding district of Fischbachau. The interior is empty, and the object of contemplation is situated outside the park. A staircase leads to a wooden bench placed in front of a round window that frames the horizon and a cross erected on the edge of the lawn opposite. The seating, a font in marble and a wooden candle-holder shelf comprise the only furniture. The construction is in a stone known to the locals as "God's concrete". The roof, doors and windows are in copper and wood.

项目名称：St. Jacob's Chapel
地点：Auerberg, Fischbachau, Germany
建筑师：architetto Michele De Lucchi S.r.l.
本地建筑师：Benno Bauer
项目团队：Marcello Biffi, Francesco Faccin, Giuseppe Filippini
面积：15m²
施工时间：2010—2012
摄影师：©Thomas Koller (courtesy of the architect)

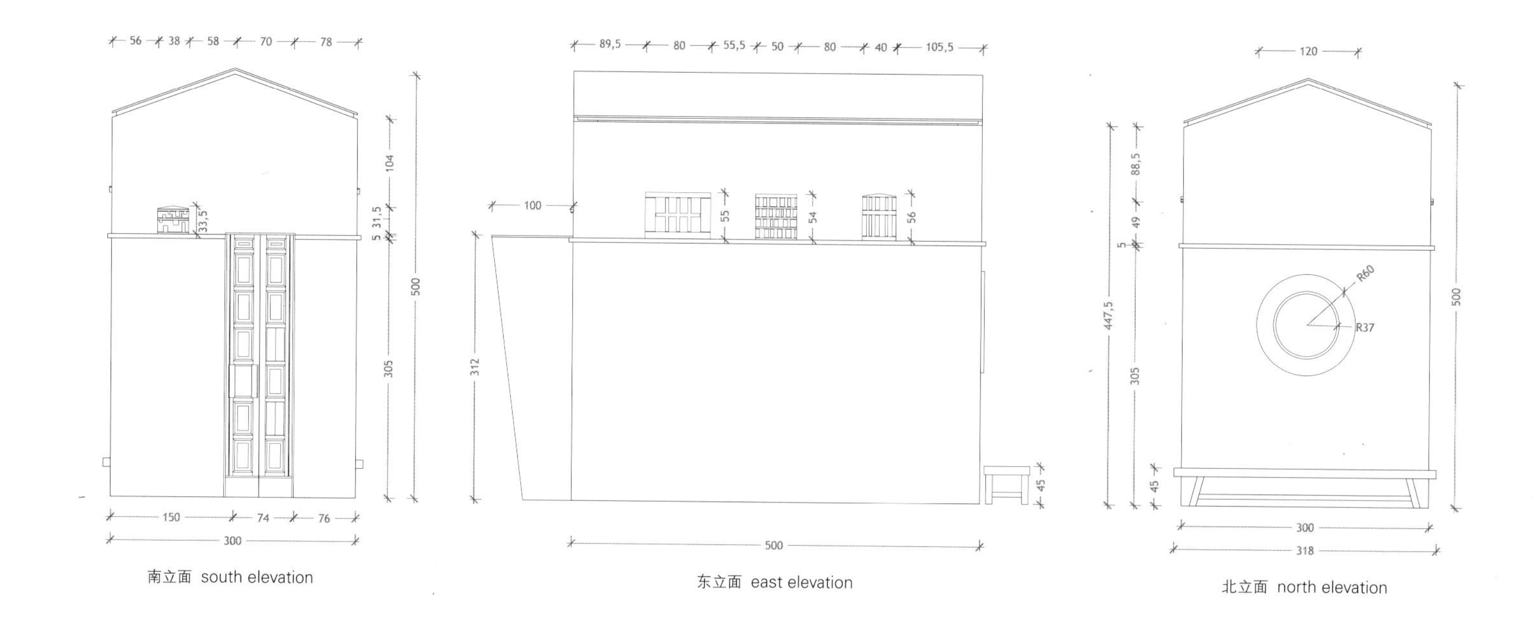

南立面 south elevation　　东立面 east elevation　　北立面 north elevation

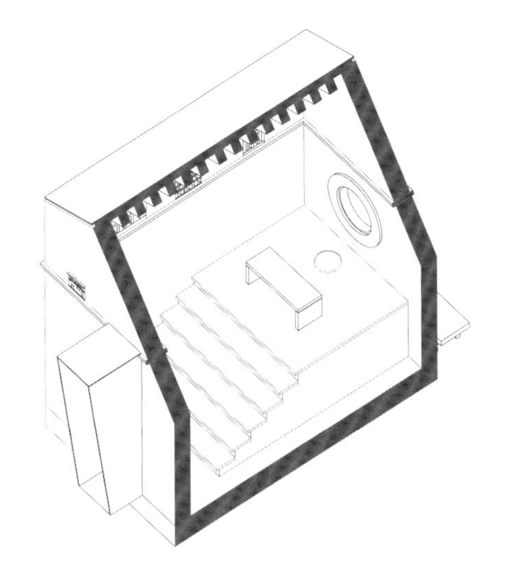

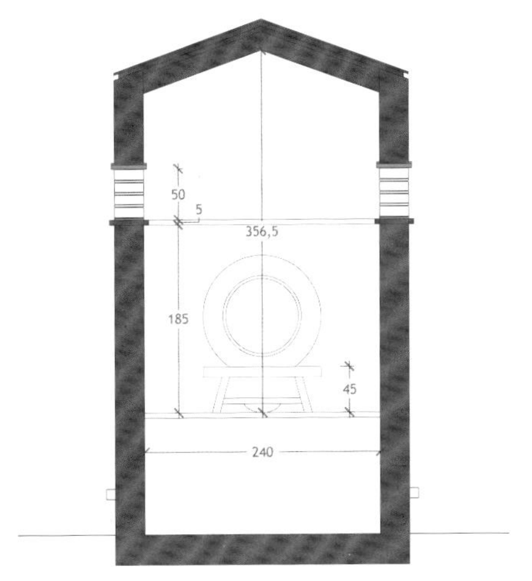

A–A' 剖面图 section A-A'

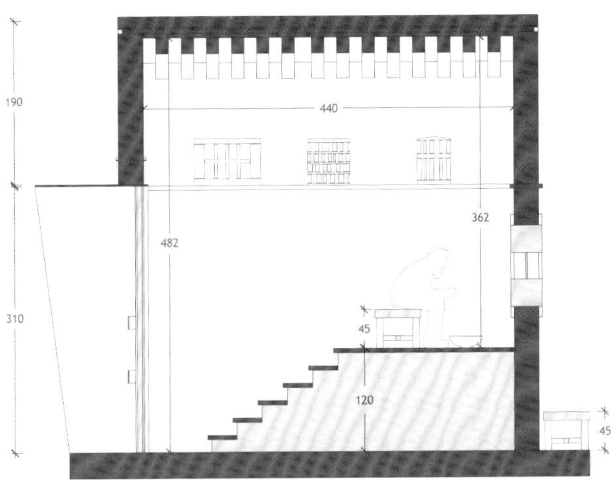

B–B' 剖面图 section B-B'

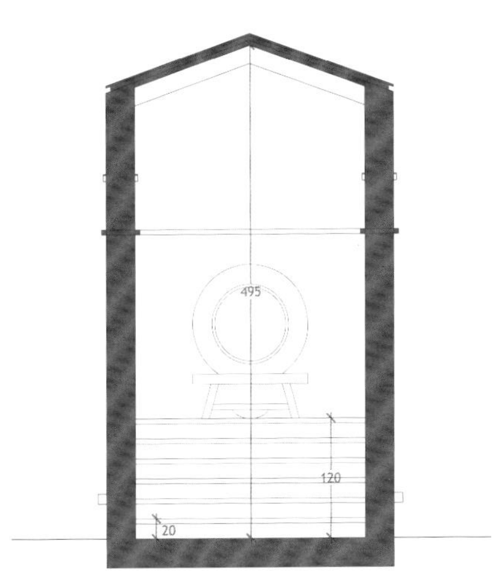

C–C' 剖面图 section C-C'

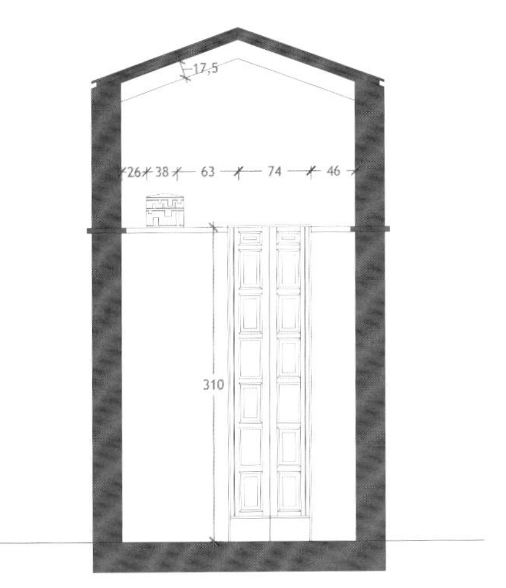

D–D' 剖面图 section D-D'

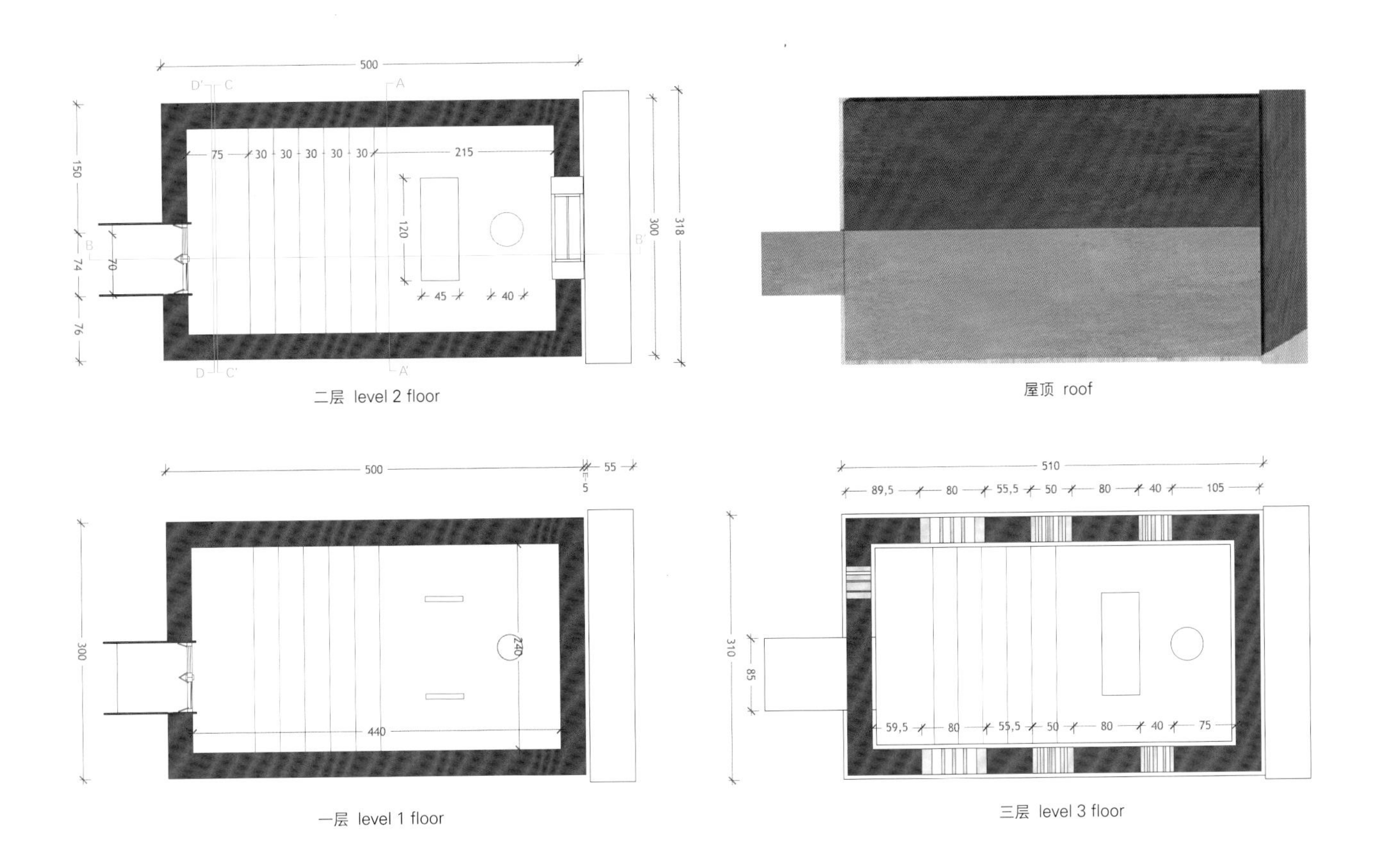

二层 level 2 floor

屋顶 roof

一层 level 1 floor

三层 level 3 floor

>>148

Yutaka Kawahara Design Studio

Yutaka Kawahara was born in 1968, Tokyo, Japan. Graduated from the Department of Architecture at Kobe University in 1984. In 2002, established Yutaka Kawahara Design Studio in Tokyo. Received numerous national and international prizes; IES Award, GPC Award, Landscape Lighting Awards and BELCA Award.

>>134

Aidlin Darling Design

Bridges the demands of artistic endeavor, environmental responsibility, functional pragmatics, and financial considerations. As a multidisciplinary firm, Aidlin Darling believes that innovations discovered through the process of design and construction can be applied to projects of any scale, use, or purpose. The studio has a broad focus including institutional, commercial and residential architecture as well as furniture and interior design. Joshua Aidlin[right] and David Darling[left] have cultivated a team that strives to deliver extraordinary, responsible and innovative design. Their approach is client and site specific, and questions conventional assumptions. In each project, they seek to uncover an inherent spirit of place and interpret constraints as catalysts for performative design. The individual characteristics of each project emerge through poetic spatial relationship, material richness, and exact detailing.

©Giovanni Gastel

>>170

Michele de Lucchi

Was born in 1951 in Ferrara, Italy. Graduated in architecture in Florence. Has designed lamps and furniture for the most known Italian and European companies; Artemide, Alias, Unifor, Hermès, Alessi. Has been work for Olivetti as director of design and developed experimental projects for Compaq Computers, Philips and Siemens. Designed and restored buildings in Japan for NTT, in Germany for Deutsche Bank, in Switzerland for Novartis, and in Italy for Enel, Piaggio, Poste Italiane, Telecom Italia. In 1999 he has been appointed to renovate some of ENEL's (the Italian Electricity Company) power plants.

>>58

DUST

Is an alliance of architects, craftsmen, artists, designers, and builders that focus their practice in the master builder tradition. Was founded in Tucson, Arizona, USA in 2007. Has grown through a collaboration between Cade Hayes[right] and Jesús Robles[left] as an exploration of ideas and ethics that had been culminating from their travels and professional experiences. Natives to the deserts of the southwestern U.S., they have cultivated an intimacy to the land in which their work is deeply informed by and attuned to, responding to the environment and the senses. Through attention and craft, the subtleties of their spirit, knowledge, and experience resonate in pursuit of the timeless qualities that remain to inspire and make us dream.

GAFPA

Was established in 2008, Gent, by Belgian architects Floris De Bruyn, Philippe De Berlangeer and Frederick Verschueren, masters in architecture at the Faculty of Architecture in KU Leuven University Gent. The firm is known for its pragmatic approach by dismantling conventional context and transforming it into new reality. GAFPA's intervention, on different levels, ranges from private houses to public buildings, from urban design operations to scenography. They regularly compete in international contests.

Níall McLaughlin Architects

Níall McLaughlin was born in Geneva in 1962, and was educated in Dublin. Received his architectural qualifications from University College Dublin in 1984. Established his own practice in London in 1990. He was chair of the RIBA Awards Group from 2007 to 2009. Was also a visiting professor of architecture at University College London as well as a visiting professor at the University of California Los Angeles from 2012 to 2013 and appointed as Lord

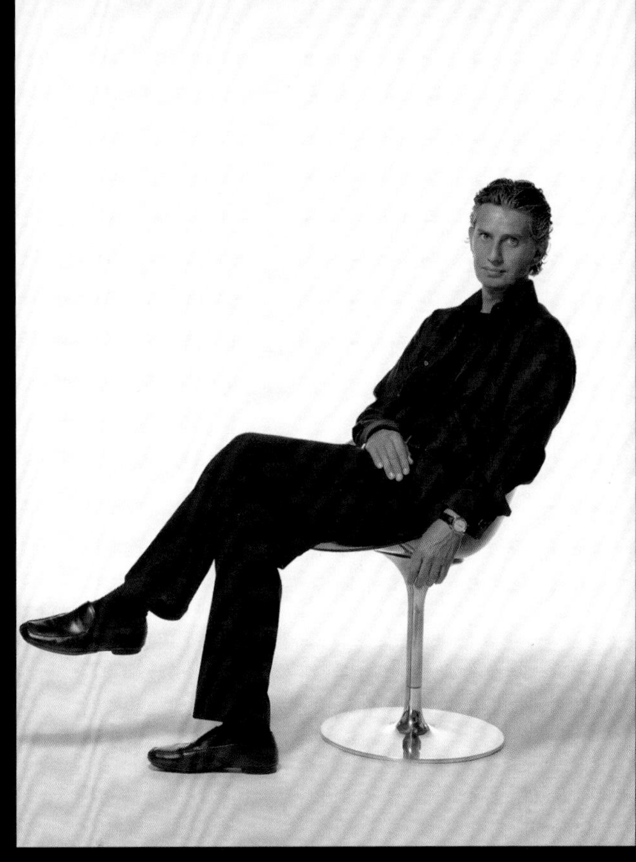

>>118

Miguel Ángel Aragonés

Was born in Mexico, in 1962. Self-taught and alien. Isolated himself by conviction from circles, schools and associations of architects. Has been energetically making original architectural work over two decades. Participated in the first international biennale of architecture, organized by the Instituto Nacional de Bellas Artes, as well as the third biennial by the National Museum of San Carlos in Mexico City. During 2002 and 2003, taught postgraduate courses in the Anáhuac University.

©Mikael Olsson

>>66

Valerio Olgiati
Was born in 1958, Chur. Studied architecture at ETH Zurich. In 1996, opened his own practice in Zurich and then in 2008 together with his wife Tamara in Flims, Switzerland. The first time he received attention was in 1999 with the museum The Yellow House in Flims. His most important buildings include the Schoolhaus in Paspels and visitor centre for the Swiss National Park in Zernez. The major solo exhibition of his work took place in 2012 at MoMa Tokyo. Since 2002 he has been a Full Professor at the Academy of architecture, Mendrisio and University of Lugano.

>>112

ADR
Is architectural studio founded by Ales Lapka and Petr Kolar in the Czech Republic in 1996. Currently with many architects and engineers, they provide a wide range of services in architecture. Specializes in constructional projects, renovations, interior design, furniture design and presenting exhibitions.

>>78

Olson Kundig Architects
Tom Kundig is a principal and owner of Seattle-based Olson Kundig Architects. Was born in 1954, California. Received undergraduate and graduate architecture degrees from the University of Washington. After working for other firms around the world, Kundig joined Olson Kundig Architects in 1986. Has received some of the world's highest design honors, from a National Design Award from the Smithsonian Cooper-Hewitt National Design Museum to an Academy Award in Architecture from the American Academy of Arts and Letters. His work has appeared in hundreds of publications worldwide including The New York Times, Architectural Record, Financial Times and The Wall Street Journal.

Douglas Murphy
Studied architecture at the Glasgow School of Art and the Royal College of Art, completing his studies in 2008. As a critic and historian, he is the author of The Architecture of Failure (Zero Books, 2009), on the legacy of 19th century iron and glass architecture, and the Last Futures (Verso, 2015), on dreams of technology and nature in the 1960s and 70s. Is also an architecture correspondent for Icon Magazine, and writes regularly for a wide range of publications on architecture and culture.

>>38

OFIS Arhitekti

Was established by Špela Videčnik and Rok Oman in 1996. OFIS project always starts with the search for a critical issue with the program, site and the client. Their design is not about surpassing, confronting, ignoring or disobeying the rules and limitations of each project. Rather, they plunge right in the middle of them and obey the 'law' literally, if need be and at times even exaggerate it. In their work, restrictions become opportunities.

>>160

BNKR Arquitectura

Is a Mexico City based architecture, urbanism and research office founded by Esteban Suarez in 2005. Has been exploring and experimenting architecture on the widest possible scale: from small iconic chapels for private clients to an urban master plan for an entire city. Believes in specific and outside the box solution to each problem, leaving behind the quest for personal style or branded architecture.

>>50

Bureau A

Founded in 2012, Bureau A is the association of Leopold Banchini[right] and Daniel Zamarbide[left]. Is a multidisciplinary platform aiming to blur the boundaries of research and project making on architectural related subjects, regardless of their nature and status. Their idea is profoundly rooted in architectural culture and history, as a vast field of exploration related to construction and installation of environments for specific purposes. Has been developing interest in public spaces and political issues; how design and architecture can possibly confront them is at the heart of the interrogations cultivated by the studio. Invests widely in research, cultivating its partner's interest in cultural exchanges and transmission through education. Taught at the Geneva University of Art and Design (HEAD). Is now guest professor at the Polytechnical School of Architecture in Lausanne (EPFL). In 2014, their work was presented at the Swiss Architecture Museum in Basel for the exhibition "Orientations: Young Swiss Architects".

>>102

Onion

Directors, Arisara Chaktranon[right] and Siriyot Chaiamnuay[left] studied architecture at the Chulalongkorn University, Bangkok. Arisara Chaktranon received Masters degree in interdisciplinary design course in Interior, Industrial and Identity at the Design Academy Eindhoven. Worked at RDG planning & Design, Bangkok and Orbit Design studio. From 2007 has been design director at Onion CO.Ltd. Siriyot Chaiamnuay received a M.Arch from AA school, London. Worked for Architect 110, Bangkok and Zaha Hadid Architects, London. Has been guest speaker at Chulalongkorn University, KMUTT(King Mongkut's University of Technology Thonburi), Chiangmai University and Kasetsart University. Was visiting instructor at the Department of Urban and Regional Planning, Chulalongkorn University.

图书在版编目(CIP)数据

休养和度假住宅 : 汉英对照 / 斯洛文尼亚OFIS arhitekti建筑师事务所等编 ; 倪琪, 栾一斐译. — 大连 : 大连理工大学出版社, 2017.3
(建筑立场系列丛书)
ISBN 978-7-5685-0748-6

Ⅰ. ①休… Ⅱ. ①斯… ②倪… ③栾… Ⅲ. ①住宅-建筑设计-汉、英 Ⅳ. ①TU241

中国版本图书馆CIP数据核字(2017)第055587号

出版发行:大连理工大学出版社
(地址:大连市软件园路80号 邮编:116023)
印 刷:上海锦良印刷厂
幅面尺寸:225mm×300mm
印 张:11.5
出版时间:2017年3月第1版
印刷时间:2017年3月第1次印刷
出 版 人:金英伟
统 筹:房 磊
责任编辑:张昕焱
封面设计:王志峰
责任校对:张媛媛
书 号:978-7-5685-0748-6
定 价:258.00元

发 行:0411-84708842
传 真:0411-84701466
E-mail:12282980@qq.com
URL: http://www.dutp.cn